Titles in This Series

Volume 4

CRM MONOGRAPH SERIES

Centre de Recherches Mathématiques
Université de Montréal

Dynamical Zeta Functions for Piecewise Monotone Maps of the Interval

David Ruelle

The Centre de Recherches Mathématiques (CRM) of the Université de Montréal was created in 1968 to promote research in pure and applied mathematics and related disciplines. Among its activities are special theme years, summer schools, workshops, postdoctoral programs, and publishing. The CRM is supported by the Université de Montréal, the Province of Québec (FCAR), and the Natural Sciences and Engineering Research Council of Canada. It is affiliated with the Institut des Sciences Mathématiques (ISM) of Montréal, whose constituent members are Concordia University, McGill University, the Université de Montréal, the Université du Québec à Montréal, and the Ecole Polytechnique.

American Mathematical Society
Providence, Rhode Island USA

The production of this volume was supported in part by the Chaire André Aisenstadt, the Fonds pour la Formation de Chercheurs et l'Aide à la Recherche (Fonds FCAR), and the Natural Sciences and Engineering Research Council of Canada (NSERC).

1991 *Mathematics Subject Classification.* Primary 58F20, 58F03; Secondary 58F11.

Library of Congress Cataloging-in-Publication Data

Ruelle, David.
    Dynamical zeta functions for piecewise monotone maps of the interval/David Ruelle.
      p. cm.—(CRM monograph series; v. 4)
    Includes bibliographical references.
    ISBN 0-8218-6991-4
    1. Differentiable dynamical systems. 2. Functions, Zeta. 3. Mappings (Mathematics)
4. Monotone operators. I. Title. II. Series.
QA614.8.R837 1994
514′.74—dc20

94-6986
CIP

Contents

Foreword

The present monograph is based on the Aisenstadt lectures given by the author in October 1993 at the Université de Montréal on "Dynamical Zeta Functions". But the emphasis is different. On one hand two excellent reviews of the subject already exist, due to Parry and Pollicott [33], and to Baladi [3]. On the other hand the theory of zeta functions for hyperbolic dynamical systems is in a state of flux because of current work by Rugh [45] and Fried. Hyperbolic systems are thus not discussed in detail here. After a general introduction (Chapter 1) we concentrate on piecewise monotone maps of the interval, and give a detailed proof of a generalized form of the theorem of Baladi and Keller [4] (Chapter 2). The Baladi-Keller theorem is typical of what one wants to prove about zeta functions associated with various kinds of dynamical systems, and the version presented here appears reasonably final.

October 1993   David Ruelle

CHAPTER 1

An Introduction to Dynamical Zeta Functions

It is not immediately obvious that the dynamical zeta functions that we shall introduce below are interesting mathematical objects. Our purpose in this Chapter 1 will be to give arguments showing that they are indeed natural and interesting objects of study. We shall at the same time introduce concepts needed for Chapter 2, making the monograph reasonably self-contained.

For more details pertaining to this Chapter 1 we refer to the monograph of Parry and Pollicott [33] and the review article by Baladi [3]; these sources also contain extensive references to the literature on dynamical zeta functions.

Note that in this Chapter 1 we have priviledged readability over strict logical organization and completeness: this is an introduction rather than a review.

1. Counting periodic orbits for maps and flows.

Let $f : M \to M$ be a map, and write Fix $f^m = \{x : f^m x = x\}$. Let $\varphi : M \to \mathcal{M}_d(\mathbb{C})$ be a matrix-valued function. If Fix f^m is finite for all m, we may define the formal power series

$$(1.1) \qquad \zeta(z) = \exp \sum_{m=1}^{\infty} \frac{z^m}{m} \sum_{x \in \text{Fix } f^m} \text{tr} \prod_{k=0}^{m-1} \varphi\big(f^k x\big).$$

Let $(f^t)_{t \geq 0}$ be a one-parameter semigroup of maps $f^t : M \to M$ (semi-flow), and $(\varphi^t)_{t \geq 0} : M \to \mathcal{M}_d(\mathbb{C})$ a family of matrix-valued functions such that $\varphi^0 = 1$ and $\varphi^{s+t}(x) = \varphi^s(f^t x)\varphi^t(x)$. We denote by P the set of periodic orbits, and by $T(\gamma)$ the period of $\gamma \in P$. If x_γ is an arbitrary point of γ, we define

$$(1.2) \qquad \zeta = \prod_{\gamma \in P} \left[\det\Big(1 - \varphi^{T(\gamma)}(x_\gamma)\Big)\right]^{-1}.$$

(We ignore convergence problems at this point).

1

In particular, given $B : M \to \mathbb{C}$ we may choose φ^t (with $d = 1$) such that $\varphi^t(x) = \exp \int_0^t \left(B(f^u x) - s \right) du$ and we find

$$\zeta = \zeta(s) = \prod_{\gamma \in P} \left[1 - \exp \int_0^{T(\gamma)} \left(B(f^t x_\gamma) - s \right) dt \right]^{-1} .$$

Taking $B = 0$ gives

$$\zeta(s) = \prod_{\gamma \in P} \left(1 - e^{-sT(\gamma)} \right)^{-1} .$$

There are obvious variations of (1.1) and (1.2) where the matrix-valued functions φ or φ^t are replaced by vector bundle maps over f or f^t.

The product formula (see Section 4 below) will show that the definitions (1.1) and (1.2) are intimately related. We shall refer loosely to objects of the type just introduced as *dynamical zeta functions*. They count periodic orbits (for maps or flows) with *weights* (associated with the function φ).

2. Subshifts of finite type.

Let I be a (nonempty) finite set, called *alphabet*. (We may take $I = \{1, \ldots, \text{card } I\}$). Let also $t = (t_{ij})$ be a matrix indexed by $I \times I$, and with elements 0 or 1, called *transition matrix*. The set I with the discrete topology is compact, and therefore the product $I^{\mathbb{Z}}$ with the product topology is also compact. A closed subset Λ of $I^{\mathbb{Z}}$ is defined by

$$\Lambda = \left\{ (\xi_k)_{k \in \mathbb{Z}} : t_{\xi_k \xi_{k+1}} = 1 \text{ for all } k \right\}$$

and a continuous map $\tau : \Lambda \to \Lambda$ is defined by

$$\left(\tau(\xi_\bullet) \right)_l = \xi_{l+1}.$$

The pair (Λ, τ) is called a *subshift of finite type*, and the map τ is called *shift*.

PROPOSITION 2.1 (Bowen-Lanford formula). *The zeta function*

$$\zeta(z) = \exp \sum_{m=1}^{\infty} \frac{z^m}{m} \text{ card Fix } \tau^m$$

extends to the rational function

$$\zeta(z) = \frac{1}{\det(1 - zt)} .$$

The basic observation is that

$$\operatorname{card} \operatorname{Fix} \tau^m = \operatorname{tr} t^m.$$

Using also the general formula

$$\det \exp A = \exp \operatorname{tr} A$$

we find

$$\zeta(z) = \exp \sum_{m=1}^{\infty} \frac{z^m}{m} \operatorname{card} \operatorname{Fix} \tau^m$$

$$= \exp \sum_{m=1}^{\infty} \frac{z^m}{m} \operatorname{tr} t^m$$

$$= \exp \operatorname{tr} \left[-\log(1 - zt) \right]$$

$$= \left[\det(1 - zt) \right]^{-1}$$

as first noted by Bowen and Lanford [**9**].

3. The product formula for maps.

If γ is a periodic orbit of (minimal) period n for $f : M \to M$, and $x_\gamma \in \gamma$, we write

$$\Phi(\gamma^q) = \operatorname{tr} \prod_{k=0}^{nq-1} \varphi\left(f^k x_\gamma\right)$$

$$= \operatorname{tr} \left[\left(\prod_{k=0}^{n-1} \varphi\left(f^k x_\gamma\right) \right)^q \right].$$

Let $\operatorname{Per}(n)$ be the set of periodic orbits of minimal period n. We then have

$$\sum_{x \in \operatorname{Fix} f^m} \operatorname{tr} \prod_{k=0}^{m-1} \varphi\left(f^k x\right) = \sum_{n \mid m} \sum_{\gamma \in \operatorname{Per}(n)} n \Phi(\gamma^{m/n})$$

where $n \mid m$ means that n divides m. Therefore

$$\zeta(z) = \exp \sum_{m=1}^{\infty} \sum_{n \mid m} \frac{n}{m} z^m \sum_{\gamma \in \operatorname{Per}(n)} \Phi(\gamma^{m/n})$$

$$= \exp \sum_{p=1}^{\infty} \sum_{\gamma \in \operatorname{Per}(p)} \sum_{q=1}^{\infty} \frac{z^{pq}}{q} \Phi(\gamma^q)$$

$$= \exp \sum_{p=1}^{\infty} \sum_{\gamma \in \operatorname{Per}(p)} \left[-\operatorname{tr} \log \left(1 - z^p \prod_{k=0}^{p-1} \varphi\left(f^k x_\gamma\right) \right) \right]$$

$$= \prod_{p=1}^{\infty} \prod_{\gamma \in \mathrm{Per}(p)} \left[\det \left(1 - z^p \prod_{k=0}^{p-1} \varphi \left(f^k x_\gamma \right) \right) \right]^{-1}.$$

We have obtained the *product formula*

$$\zeta(z) = \prod_{p=1}^{\infty} \prod_{\gamma \in \mathrm{Per}(p)} \left[\det \left(1 - z^p \prod_{k=0}^{p-1} \varphi \left(f^k x_\gamma \right) \right) \right]^{-1}.$$

If we write $p(\gamma)$ for the (minimal) period of γ, this is:

$$\zeta(z) = \prod_{\gamma \in P} \left[\det \left(1 - z^{p(\gamma)} \prod_{k=0}^{p(\gamma)-1} \varphi \left(f^k x_\gamma \right) \right) \right]^{-1}.$$

In particular, when φ is the constant scalar function 1 we have

$$\zeta(z) = \exp \sum_{m=1}^{\infty} \frac{z^m}{m} \, \mathrm{card} \, \mathrm{Fix} \, f^m$$

$$= \prod_{p=1}^{\infty} (1 - z^p)^{- \, \mathrm{card} \, \mathrm{Per}(p)}.$$

Note that, at this point, all the formulae hold only at the level of formal power series.

EXAMPLE 3.1. The map $x \mapsto 1 - \mu x^2$ of the interval $[-1, 1]$ to itself for the Feigenbaum value $\mu = 1.401155\ldots$ has one periodic orbit of period 2^n for each integer $n \geq 0$. Therefore

$$\zeta(z) = \exp \sum_{n=0}^{\infty} \frac{2^{n+1} - 1}{2^n} z^{2^n}$$

$$= \prod_{n=0}^{\infty} \left(1 - z^{2^n} \right)^{-1}$$

$$= \prod_{n=0}^{\infty} \left(1 + z^{2^n} \right)^n$$

where the last step used the identity

$$(1 - z)^{-1} = \prod_{n=0}^{\infty} \left(1 + z^{2^n} \right).$$

4. The product formula for semiflows.

Suppose that the semiflow (f^t) has a global section Σ. Every orbit $(f^t x)_{t \geq 0}$ thus intersects Σ, and there is a function $t : \Sigma \to \mathbb{R}$ such that

$t(x)$ is the smallest real > 0 with the property $f^{t(x)}x \in \Sigma$. We define $f : \Sigma \to \Sigma$ by $fx = f^{t(x)}x$, and write $\varphi(x) = \varphi^{t(x)}(x)$. Then

$$\zeta = \prod_{\gamma \in P} \left[\det \left(1 - \varphi^{t(\gamma)}(x_\gamma) \right) \right]^{-1}$$

$$= \prod_{p=1}^{\infty} \prod_{\gamma \in \mathrm{Per}(p)} \left[\det \left(1 - \prod_{k=0}^{p-1} \varphi\left(f^k x_\gamma\right) \right) \right]^{-1}$$

$$= \zeta(z)\Big|_{z=1}.$$

This formula (where convergence questions have been ignored) relates the zeta function for a semiflow and the zeta function for a map.

5. The Lefschetz formula.

If M is a compact manifold, one can define the index $L(x, f) \in \mathbb{Z}$ of an isolated fixed point x for a continuous map $f : M \to M$. If f is differentiable at x, and $1 - D_x f$ is invertible, then $L(x, f) = \mathrm{sign}\det(1 - D_x f)$. The sum $\sum_{x \in \mathrm{Fix}\, f} L(x, f)$ (defined when $\mathrm{Fix}\, f$ is finite) is a homotopy invariant.

The *Lefschetz number* of f is defined by

$$\Lambda(f) = \sum_{i=0}^{\dim M} (-1)^i \,\mathrm{tr}\, f_{*i}$$

where f_{*i} is the automorphism induced by f on the *i-th* (singular) homology group $H_i(M, Q)$ with rational coefficients. We have then the *Lefschetz trace formula*

$$\sum_{x \in \mathrm{Fix}\, f} L(x, f) = \Lambda(f).$$

Define a Lefschetz zeta function

$$\tilde{\zeta}(z) = \exp \sum_{m=1}^{\infty} \frac{z^m}{m} \sum_{x \in \mathrm{Fix}\, f^m} L(x, f^m)$$

(where we suppose that $\mathrm{Fix}\, f^m$ is finite for all m). Then

$$\tilde{\zeta}(z) = \exp \sum_{m=1}^{\infty} \frac{z^m}{m} \sum_i (-1)^i \,\mathrm{tr}\, f_{*i}^m$$

$$= \prod_i \left[\exp \mathrm{tr} \sum_{m=1}^{\infty} \frac{z^m}{m} f_{*i}^m \right]^{(-1)^i}$$

$$= \prod_i \left[\det(1 - z f_{*i}) \right]^{(-1)^{i+1}}.$$

This function is therefore rational, and has a homological interpretation.

Let x be hyperbolic with minimal period p ($D_x f^p$ has no eigenvalue λ with $|\lambda| = 1$), and let E^u be the subspace corresponding to the eigenvalues λ of $D_x f^p$ with $|\lambda| > 1$. Denoting by $\gamma = \{x, \ldots, f^{p-1}x\}$ the orbit of x we write $u(\gamma) = \dim E^u$ and $\Delta(\gamma) = \pm 1$ depending on whether Df^p preserves or reverses the orientation of E^u. Following Smale [46] we remark that

$$L(x, f^{pq}) = (-1)^{u(\gamma)} \Delta(\gamma)^q.$$

Therefore if all periodic points of f are hyperbolic, we have a product formula

$$\tilde{\zeta}(z) = \exp \sum_p \sum_{\gamma \in \mathrm{Per}(p)} (-1)^{u(\gamma)} \sum_q \frac{z^{pq}}{q} \Delta(\gamma)^q$$

$$= \prod_p \prod_{\gamma \in \mathrm{Per}(p)} \left[1 - \Delta(\gamma) z^p \right]^{(-1)^{u(\gamma)+1}}$$

$$= \prod_{\gamma \in P} \left[1 - \Delta(\gamma) z^{p(\gamma)} \right]^{(-1)^{u(\gamma)+1}}.$$

Note that in the holomorphic case $\Delta(\gamma) = 1$, $u(\gamma)$ is even, and $\tilde{\zeta}(z)$ reduces to $\zeta(z)$.

A natural idea is now to introduce dynamical Lefschetz zeta functions of the type

$$\tilde{\zeta}(z) = \exp \sum_{m=1}^{\infty} \frac{z^m}{m} \sum_{x \in \mathrm{Fix}\, f^m} L(x, f^m) \, \mathrm{tr} \prod_{k=0}^{m-1} \varphi\big(f^k x\big).$$

In general one will try to reduce the study of dynamical zeta functions to that of Lefschetz zeta functions (the latter are in some sense more natural and certainly easier to analyze).

In order to proceed, and in particular to study the convergence of the formal power series $\zeta(z)$, we shall have to make specific choices for the dynamical system considered and the functional space of the weight function φ. It turns out that some rather different choices are possible and interesting. But this also means that the theory of dynamical zeta functions tends to split up into a number of specialized branches: all are related but different, and a unifying theory is missing. (Chapter 2 of this monograph will explore in detail one of the "specialized branches", where f is assumed to be a piecewise monotone map of the interval). To gain perspective on the subject, it is convenient at this point to digress on the history of zeta functions.

6. Historical note: From the Riemann zeta function to dynamical zeta functions.

We may write (for real $s > 1$)

$$\zeta(s) \equiv \sum_{n=0}^{\infty} \frac{1}{n^s} = \prod_{p \text{ prime}} (1 - p^{-s})^{-1}.$$

This product formula was discovered by Euler (18-th century), but the detailed analytic study of ζ (aimed at number-theoretic applications) is due to Riemann (19-th century) hence the name of *Riemann zeta function*.

Given an integer $m > 0$, the residue classes (n) mod m with $(n, m) = 1$ form a multiplicative group. Let χ be a *character* of that group, and write $\chi(n) = 0$ if $(n, m) \neq 1$. The *Dirichlet L-function* is defined by

$$L(s, \chi) \equiv \sum_{n=1}^{\infty} \frac{\chi(n)}{n^s} = \prod_{p} \left(1 - \chi(p)p^{-s}\right)^{-1}.$$

Various other functions similar to the Riemann zeta function and Dirichlet L-function have been introduced later, often in view of number theoretic applications, and with typically the following properties[1]:

   (i) Meromorphy in the whole complex plane. [The position of poles and zeros has been a prime object of study: the Riemann zeta function has a simple pole at $s = 1$, simple zeros at $s = -2, -4, \ldots, -2n, \ldots$ (trivial zeros), and the Riemann hypothesis asserts that all other zeros are on the line $\operatorname{Re} s = \frac{1}{2}$ (nontrivial zeros)].
   (ii) Dirichlet series expansion $\sum_n a_n e^{-\lambda_n s}$.
   (iii) Euler product expansion.
   (iv) Functional equation. [For the Riemann zeta function, if we write $\xi(s) = \pi^{-s/2}\Gamma\left(\frac{s}{2}\right)\zeta(s)$, the functional equation is $\xi(s) = \xi(1 - s)$].

Let k be a finite field with q elements and V a projective nonsingular algebraic variety of dimension n defined over k. [Note that the points of V have coordinates in the algebraic closure of k, but the defining equations have coefficients in k]. If N_m is the number of points of V with coordinates in the extension field of degree m of k, one defines the

---

[1]For a general discussion see the article "Zeta functions" in the Encyclopedic Dictionary of Mathematics (Nihon Sugakkai [**30**]).

zeta function of V as

$$Z(z, V) = \exp \sum_{m=1}^{\infty} N_m \frac{z^m}{m}.$$

Note that

$$N_m = \text{card Fix } F^m$$

where F is the *Frobenius morphism*, which replaces the point of coordinates (x_i) by the point (x_i^q). Certain conjectures proposed by Weil on the properties of $Z(z, V)$ led to a lot of work by Weil, Dwork, Grothendieck and others, and a complete proof was finally obtained by Deligne. It is found that $Z(z, V)$ is a rational function of z:

$$Z(z, V) = \prod_{l=0}^{2n} P_l(z)^{(-1)^{l+1}}$$

where the zeros of the polynomial P_l have absolute value $q^{-l/2}$ and the P_l have a cohomological interpretation. [Roughly, $P_l(z) = \det\left(1 - zF^*|H^l(V)\right)$ where F is the action of the Frobenius morphism on cohomology].

If the Frobenius morphism of the algebraic variety V is replaced by a diffeomorphism f of a smooth compact manifold, one obtains the definition

(6.1) $$\zeta(z) = \exp \sum_{m=1}^{\infty} \frac{z^m}{m} \text{ card Fix } f^m.$$

Artin and Mazur [1] have shown that, for a C^1-dense set of diffeomorphisms,

$$\limsup_{m \to \infty} \frac{1}{m} \log \text{card Fix } f^m < \infty$$

and therefore the zeta function (6.1) has a strictly positive radius of convergence. Smale [46] then conjectured that, for the Axiom A diffeomorphisms, which he had introduced, the Artin-Mazur zeta function is rational. This was later proved by Guckenheimer [17] and Manning [25] (also Bowen [8], Fried [13]).

Let $\{\gamma\}$ be the set of closed geodesics on a compact surface M of constant negative curvature -1, and $l(\gamma)$ the length of γ. The *Selberg zeta function* is defined by

(6.2) $$Z(s) = \prod_{\gamma} \prod_{k=0}^{\infty} \left(1 - e^{-(s+k)l(\gamma)}\right).$$

It is an entire function of order 2 and satisfies a functional equation. It has "trivial zeros" at $0, -1, \ldots, -n, \ldots$ and "nontrivial zeros" with

$\operatorname{Re} s = \frac{1}{2}$, except for a finite number on the interval $(0, 1)$. (The non-trivial zeros of Z are related to the eigenvalues of the Laplace operator on M).

One may interpret $l(\gamma)$ as the period of a periodic orbit for the geodesic flow on M. This suggests defining a zeta function for the flow (f^t) by

$$(6.3) \qquad \zeta(s) = \prod_{\gamma \in P} \left(1 - e^{-sT(\gamma)}\right)^{-1}$$

where P is the set of periodic orbits, and $T(\gamma)$ is the period of γ. [In the case of the geodesic flow on a compact surface of curvature -1, we have $\zeta(s) = Z(s+1)/Z(s)$. Smale [46] proposed to define the zeta function of a flow by (6.2), but this definition is reasonable only for the geodesic flow with curvature -1; in general (6.2) does not behave well under changes of time scale].

As we have seen, the consideration of *arithmetic zeta functions* leads naturally to the definitions (6.1) and (6.3) of zeta functions counting periodic orbits of dynamical systems. The ideas of equilibrium statistical mechanics suggest however to count periodic orbits with weights, i.e., to replace (6.1), (6.3) by

$$(6.4) \qquad \zeta(z) = \exp \sum_{m=1}^{\infty} \frac{z^m}{m} \sum_{x \in \operatorname{Fix} f^m} \exp \sum_{k=0}^{m-1} A\left(f^k x\right)$$

or

$$(6.5) \qquad \zeta(s) = \prod_{\gamma \in P} \left[1 - \exp\left(-\int_0^{T(\gamma)} \left(s - B(f^t x_\gamma)\right) dt\right)\right]^{-1}$$

where $x_\gamma \in \gamma$. These formulae define *dynamical zeta functions*.

Note that the trivial choice $B = 0$ in (6.5), which reproduces (6.3), corresponds by the product formula of Section 4 to a non-trivial choice of A in (6.4). This makes the introduction of a weight $\varphi = e^A$ very natural. The definitions (6.4) and (6.5) were introduced and studied by Ruelle [36, 37, 38] (in the case of Axiom A dynamical systems).

It turns out that the dynamical zeta functions are closely related to problems of ergodic theory (decay of correlations, thermodynamic formalism).

7. Properties of dynamical zeta functions.

If we compare the properties of number-theoretic zeta functions and dynamical zeta functions, it can be seen that the latter have

(i) analyticity properties that can be analyzed in detail,

(ii) Dirichlet series expansion in the semi-flow case,

(iii) a product formula,

(iv) perhaps something like a functional equation (see Ruelle [**44**]).

The parallelism is thus striking. If however we look at the (co)homological interpretation of the Lefschetz zeta function (Section 5) we see that it is completely spoilt by the introduction of weights. What happens is that instead of being able to express the zeta function in terms of the action of the dynamical system on finite dimensional cohomology groups, we have it only in terms of the action of the dynamical system on infinite dimensional cochain groups.

Let us mention at this point that Atiyah and Bott [**2**], in a classical paper, have analyzed situations where it is possible to "pass to the quotient" and reach the level of cohomology groups.

The action of the dynamical system on cochain groups referred to above is given by so-called *transfer operators*, and dynamical zeta functions will be expressed in terms of determinants of transfer operators. In some cases, these determinants will simply be Fredholm determinants (in the sense of Grothendieck, see below, Section 11). But in other cases the theory of Fredholm-Grothendieck will have to be generalized.

It is remarkable that Dwork, at an early date, has used transfer operators in a p-adic setting to study the zeta functions of algebraic hypersurfaces over a finite field. His study is analogous to the later studies made in a Hölder, differentiable, or analytic setting.

8. Transfer operators.

As earlier we consider a map $f : M \to M$, but we replace the matrix-valued function φ by a scalar function $g : M \to \mathbb{C}$. (Matrix-valued functions will reappear in a minute). We define the *transfer operator* $\mathcal{L}$ by

$$\mathcal{L}\Phi(x) = \sum_{y:fy=x} g(y)\Phi(y)$$

acting on functions $\Phi : M \to \mathbb{C}$. Note the important property

$$\mathcal{L}\Big(\Phi \cdot (\Phi' \circ f)\Big) = \Phi' \cdot (\mathcal{L}\Phi).$$

The interesting situation is of course when f is not invertible but has a finite (or at least discrete) set of inverse branches ψ_ω. (Often one can usefully convert a problem with invertible f into a problem with noninvertible f). One can rewrite $\mathcal{L}$ in terms of the inverse branches

ψ_ω, or more generally define a (generalized) transfer operator $\mathcal{K}$ by

$$\mathcal{K}\Phi(x) = \sum_\omega \varphi_\omega(x)\Phi(\psi_\omega x)$$

or

$$= \int m(d\omega)\varphi_\omega(x)\Phi(\psi_\omega x)$$

where the ψ_ω are homeomorphisms of subsets of M to subsets of M, and $m(d\omega)$ is a measure. The operator $\mathcal{K}$ acts on a Banach space B of functions on M (continuous functions usually), or more generally on a space of sections of a vector bundle.

If M and f are smooth, we may replace g by the "matrix-valued" function $g \cdot \bigwedge^l(T^*\psi_\omega)$, we obtain thus a transfer operator $\mathcal{L}^{(l)}$ acting on l-forms for $l = 0, \ldots, \dim M$, and $\mathcal{L}^{(0)}$ is the original transfer operator $\mathcal{L}$.

9. Traces and determinants.

If M and f are smooth and the graph of f is transversal to the diagonal $\Delta \subset M \times M$, it is natural to define a "trace"[2]

$$\text{Tr}\,\mathcal{L} = \sum_{x \in \text{Fix}\,f} \frac{g(x)}{|\det(1 - D_x f^{-1})|}$$

where $D_x f$ is the derivative of f acting in the tangent space $T_x M$. This is a natural definition because, using a local chart near the fixed point x, we have

$$\int \mathcal{L}(\xi, \eta)\delta(\xi - \eta)\, d\xi\, d\eta = \int g(\eta)\delta\big(\eta - f^{-1}\xi\big)\delta(\xi - \eta)\, d\xi\, d\eta$$

$$= \int g(\xi)\delta\big(\xi - f^{-1}\xi\big)\, d\xi$$

$$= g(x)/|\det\big(1 - D_x f^{-1}\big)|$$

according to distribution theory (we have defined the kernel $\mathcal{L}(\xi, \eta)$ so that $\mathcal{L}\Phi(\xi) = \int \mathcal{L}(\xi, \eta)\Phi(\eta)\, d\eta$).

The above definition extends naturally to the transfer operators $\mathcal{L}^{(l)}$:

$$\text{Tr}\,\mathcal{L}^{(l)} = \sum_{x \in \text{Fix}\,f} \frac{g(x)\,\text{tr}\big(\bigwedge^l D_x f^{-1}\big)}{|\det\big(1 - D_x f^{-1}\big)|}$$

---

[2]or "flat trace", see Atiyah and Bott [2], Guillemin and Sternberg [18].

(where tr denotes the trace of an operator in the finite dimensional space $\bigwedge^l T_x M$). We have then the miraculous result

$$\sum_{l=0}^{\dim M} (-1)^l \operatorname{Tr} \mathcal{L}^{(l)} = \sum_{x \in \operatorname{Fix} f} g(x) \frac{\det\left(1 - D_x f^{-1}\right)}{\left|\det\left(1 - D_x f^{-1}\right)\right|}$$

$$= \sum_{x \in \operatorname{Fix} f} g(x) L\left(x, f^{-1}\right).$$

(The Lefschetz index $L(x, f^{-1})$ is simply related to $L(x, f)$ and will in many cases be 1).

In terms of the "traces" Tr, let us now define a "determinant" Det by the usual formula

$$\operatorname{Det}(1 - z\mathcal{L}) = \exp - \sum_{m=1}^{\infty} \frac{z^m}{m} \operatorname{Tr} \mathcal{L}^m.$$

Then we have

$$\zeta^{\times}(z) = \prod_{l=0}^{\dim M} \left(\operatorname{Det}\left(1 - z\mathcal{L}^{(l)}\right)\right)^{(-1)^{l+1}}$$

where $\zeta^{\times}$ is a Lefschetz zeta function.

The difference between $\zeta^{\times}$ and ζ is not a serious worry (the two functions are often simply related). The serious problem is to make sense of the $\operatorname{Det}(1 - z\mathcal{L}^{(l)})$ as analytic functions in a reasonably large domain, and not just as formal power series. This will involve the spectral theory of the transfer operators and depend on the class of dynamical systems and functional spaces considered.

We shall now turn for a while to the theory of Fredholm determinants which give one well understood example of functional determinants of the type $\operatorname{Det}(1 - z\mathcal{L})$.

10. Entire analytic functions.

It is convenient to recall here some results about entire analytic functions.

Let the entire analytic function $f(z)$ vanish of order m at 0, and let (α_k) be the sequence of the further zeros repeated according to multiplicity and arranged by increasing modulus. If $\lambda > 0$ we write

$$S(\lambda) = \sum |\alpha_k|^{-\lambda}$$

and define the *exponent of convergence* ρ_0 of the zeros by

$$\rho_0 = \inf\{\lambda : S(\lambda) < +\infty\}.$$

If ρ_0 is finite we define the genus p to be the least integer ≥ 0 such that $S(p+1) < +\infty$. [Therefore $p = [\rho_0]$ except that, when ρ_0 is an integer ≥ 1, we may have $p = \rho_0 - 1$]. We may then write, as a special case of the Weierstraß product formula,

$$(10.1) \qquad f(z) = z^m e^{g(z)} \prod_k \left(1 - \frac{z}{\alpha_k}\right) \exp g_p\left(\frac{z}{\alpha_k}\right)$$

where $g_0 = 0$, $g_p(z) = \sum_{k=1}^p \frac{z^k}{k}$, and g is an entire function.

The *order* ρ of the entire analytic function $f(z) = \sum a_k z^k$ is defined by

$$\rho = \limsup_{r \to \infty} \frac{1}{\log r} \log \log \max_{|z|=r} |f(z)|$$

$$= \limsup_{n \to \infty} \frac{n \log n}{\left|\log |a_n|\right|}.$$

THEOREM 10.1. *We have*

$$\rho_0 \leq \rho.$$

If ρ is finite, then g in (10.1) is a polynomial of degree $\leq \rho$ and $\rho_0 = \rho$ unless ρ is an integer ≥ 1.

11. The theory of Fredholm-Grothendieck.

(See Grothendieck [**15, 16**]).

If $E_1, \ldots, E_n$ are Banach spaces, a norm $\| \cdot \|_1$ on $\otimes E_i$ is defined by

$$\|u\|_1 = \inf \sum_i \|x_{i1}\| \cdot \|x_{i2}\| \cdot \cdots \cdot \|x_{in}\|$$

where the infimum is taken over all representations

$$u = \sum_i x_{i1} \otimes x_{i2} \otimes \cdots \otimes x_{in}.$$

The completion of $\otimes E_i$ with respect to $\| \cdot \|_1$ is a Banach space $\widehat{\otimes} E_i$ (the *projective tensor product* of the E_i). The elements of this space are called *Fredholm kernels* by Grothendieck.

If E, F are Banach spaces, and E' denotes the dual of E, there is a canonical map $E' \widehat{\otimes} F \to \mathcal{L}(E, F)$, with obvious definition. This map is norm-reducing[3]. If $u \in E' \widehat{\otimes} F$, its image $\tilde{u} \in \mathcal{L}(E, F)$ is called a *trace-class* or *nuclear* operator $E \to F$, such operators are compact.

---

[3]The map $E' \widehat{\otimes} F \to \mathcal{L}(E, F)$ is often injective, and $E' \widehat{\otimes} F$ can then be identified with a subspace of $\mathcal{L}(E, F)$.

In particular, the image of $E'\widehat{\otimes}E$ is an ideal of $\mathcal{L}(E)$. An element $u \in E\widehat{\otimes}F$ can be written as a convergent series

$$u = \sum_{i=1}^{\infty} \lambda_i\, x_i \otimes y_i$$

where $x_i \in E$, $y_i \in F$, $\|x_i\| \leq 1$, $\|y_i\| \leq 1$, $\lambda_i \geq 0$ and $\sum \lambda_i \leq \|u\|_1 + \varepsilon$ for arbitrarity small $\varepsilon > 0$.

If we adjoin 1 to the algebra $E'\widehat{\otimes}E$ (when E is infinite dimensional), and take

$$u = \sum_{i=1}^{\infty} x_i' \otimes x_i$$

with $x_i \in E$, $x_i' \in E'$, it is natural to define

$$(11.1) \qquad \mathrm{Det}(1 + u) = 1 + \sum_{n=1}^{\infty}\ \sum_{i_1 < \cdots < i_n} \det\big(\langle x_{i_l}', x_{i_k}\rangle\big)$$

$$(11.2) \qquad\qquad = 1 + \sum_{n=1}^{\infty} \frac{1}{n!}\alpha_n(u)$$

where $\big(\langle x_{i_l}', x_{i_k}\rangle\big)$ denotes an $n \times n$ matrix with elements indexed by $k, l = 1, \ldots, n$.

If $u_1, \ldots, u_n \in E' \otimes E$, we may write

$$u_k = \sum_{i=1}^{\infty} x_i' \otimes x_{ki}$$

for $k = 1, \ldots n$, and define an n-linear symmetric function $\widehat{\alpha}_n : (E'\widehat{\otimes}E)^n \to \mathbb{C}$ by

$$\widehat{\alpha}_n(u_1, \ldots, u_n) = \sum_{i_1} \cdots \sum_{i_n} \det\big(\langle x_{i_l}', x_{ki_k}\rangle\big).$$

It is no restriction to take $\|x_i'\| = 1$, hence

$$|\widehat{\alpha}_n(u_1, \ldots, u_n)| \leq \sum_{i_1} \cdots \sum_{i_n} \prod_{k=1}^{n} \big(n^{1/2}\|x_{ki_k}\|\big)$$

$$= n^{n/2} \prod_{k=1}^{n} \sum_{i} \|x_{ki}\|$$

so that $|\widehat{\alpha}_n(u_1, \ldots, u_n)| \leq n^{n/2}\|u_1\|_1 \ldots \|u_n\|_1$. We see that (11.2) defines an entire analytic function $u \to \mathrm{Det}(1 + u)$ on $E'\widehat{\otimes}E$ since $\alpha_n(u) = \widehat{\alpha}_n(u, \ldots, u)$.

THEOREM 11.1. *The following conditions are equivalent:*
(a) *the operator $1 - \tilde{u}$ is invertible in $\mathcal{L}(E)$,*

(b) $1 - u$ is invertible in the algebra obtained by adjoining 1 to $E' \hat{\otimes} E$,

(c) $\text{Det}(1 - u) \neq 0$.

One may then write

$$(1 - u)^{-1} = \frac{R(u)}{\text{Det}(1 - u)}$$

where $R(u)$ is an operator-valued entire analytic function of u (of which the coefficients may be explicitly specified).

THEOREM 11.2. *If λ is an eigenvalue of multiplicity n of the operator $\tilde{u}$ defined by $u \in E' \hat{\otimes} E$ (i.e., if n is the dimension of the corresponding generalized eigenspace), then λ^{-1} is a zero of order exactly n of the entire function $z \mapsto \text{Det}(1 - zu)$.*

Remarks 11.3. (a) The trace $\sum x_i' \otimes x_i \mapsto \sum \langle x_i', x_i \rangle$ extends to a continuous linear form $\text{Tr} : E' \hat{\otimes} E \to \mathbb{C}$, and we have the identity

$$\text{Det}(1 - zu) = \exp - \sum_{n=1}^{\infty} \frac{z^n}{n} \, \text{Tr} \, u^n$$

between power series converging in a neighborhood of 0. In particular, if $z \mapsto \text{Det}(1 - zu)$ has order < 1, hence genus 0, then $\sum |\lambda_i| < \infty$ (where the λ_i are the eigenvalues of $\tilde{u}$) and $\text{Tr} \, u = \sum \lambda_i$.

(b) By our estimates of the coefficients of $z \mapsto \text{Det}(1 - zu)$, this entire function is of order ≤ 2, and Grothendieck [**15**, Ch. 2, p. 18] shows that

$$\text{Det}(1 - zu) = e^{-z \, \text{Tr} \, u} \prod_i (1 - z\lambda_i) e^{z\lambda_i}$$

where the product is over the non-zero eigenvalues λ_i (repeated, as always, according to multiplicity). If $\sum_i |\lambda_i| < \infty$, then

$$\text{Det}(1 - zu) = e^{-\alpha z} \prod_i (1 - z\lambda_i)$$

where $\alpha = \text{Tr} \, u - \sum_i \lambda_i$.

(c) If M and L are compact spaces and m is a measure on L, a continuous kernel $K : M \times L \to \mathbb{C}$ defines an element $u \in \mathcal{C}(L)' \hat{\otimes} \mathcal{C}(M)$, corresponding to the operator

$$\tilde{u} : \Phi \mapsto \int K(\cdot, y) \Phi(y) m(dy)$$

and we have

$$\|u\|_1 = \int \max_x |K(x, y)| \, |m(dy)|.$$

This situation covers the classical case considered by Fredholm, to which Grothendieck's theory therefore applies. Grothendieck also gives several non classical examples where the same theory applies.

12. Analyticity improving linear maps.

Let V be a d-dimensional complex manifold, and m a bounded measure with continuous density > 0 on V. For open $U \subset V$ we denote by $E(U)$ the subspace of holomorphic functions in $L^2(U, m|U)$, and by $\| \cdot \|_U$ the corresponding norm. (More generally we could let $E(U)$ be the Hilbert space of square integrable holomorphic sections of a holomorphic vector bundle over U).

PROPOSITION 12.1. *Let F be a Banach space and $\tilde{u} : E(V) \to F$ a linear map such that*

$$\|\tilde{u}\Phi\| \leq \text{const.}\|\Phi\|_U$$

where U is open with compact closure $\overline{U} \subset V$. Then $\tilde{u}$ is associated with $u \in E(V)' \widehat{\otimes} F$ and one can write $u = \sum_{k=1}^{\infty} \lambda_k x_k' \otimes y_k$ where $\|x_k'\| \leq 1$, $\|y_k\| \leq 1$ and $0 < \lambda_k \leq \exp(A - Bk^{1/d})$ with real constants A, B such that $B > 0$.

We may choose open sets W_1, W_2 with compact closures $\overline{W}_1$, $\overline{W}_2$ such that

$$\overline{U} \subset W_1 \subset \overline{W}_1 \subset W_2 \subset \overline{W}_2 \subset V.$$

Using the Cauchy formula we see that the restriction maps $E(W_2) \to E(W_1) \to E(U)$ are defined by bounded continuous kernels. Since these are square integrable they correspond to elements $\sum \lambda_i v_i' \otimes v_i$ with $\sum |\lambda_i|^2 < \infty$, $\|v_i'\| \leq 1$, $\|v_i\| \leq 1$. By composition, the restriction map $E(W_2) \to E(U)$ corresponds therefore to $\sum \lambda_i w_i' \otimes w_i$ with $\sum |\lambda_i| < \infty$, $\|w_i'\| \leq 1$, $\|w_i\| \leq 1$. The construction gives $w_i \in L^2(U, m|U)$ but by projection we may take $w_i \in E(U)$.

For $\Phi \in E(V)$, we may estimate the $w_i'(\Phi)$ in terms of the Taylor expansion of Φ at finitely many points a_j and the derivatives at the a_j in terms of Cauchy integrals. This gives

$$w_i'(\Phi) = \sum_j \sum_{k_1, \dots k_d \geq 0} \alpha^{k_1 + \dots + k_d} A_{ijk_1 \dots k_d} u_{jk_1 \dots k_d}'(\Phi)$$

where $0 < \alpha < 1$ and

$$|A_{ijk_1 \dots k_d}| \leq \text{const.}, \quad \|u_{jk_1 \dots k_d}'\| \leq \text{const.}$$

By assumption, $\tilde{u}$ is obtained by composing the restriction map $E(V) \to E(U)$ and a bounded linear map $\varphi : E(U) \to F$. Finally, we see that $\tilde{u}$ is associated with

$$\sum_j \sum_{k_1,\ldots,k_d \geq 0} \alpha^{k_1+\cdots+k_d} u'_{jk_1\ldots k_d} \otimes \sum_i \lambda_i A_{ijk_1\ldots k_d}(\varphi w_i)$$

which is of the desired form. Note that α can be chosen arbitrarily small > 0, and therefore B arbitrarily large.

COROLLARY 12.2. *Take $F = E(V)$, which means that $\tilde{u}$ is analyticity improving, then the entire function $z \mapsto \mathrm{Det}(1 - zu)$ is of order 0, and more precisely there are constants C, D such that*

$$|\mathrm{Det}(1 - zu)| \leq \exp\left(C + D\left(\log_+ |z|\right)^{d+1}\right)$$

see [**12**, Lemma 6] .

EXAMPLE 12.3. Let $\psi : V \to U$ be holomorphic, and $\varphi \in E(V)$. We define an operator on square integrable holomorphic l-forms on V by

$$(\mathcal{L}\Phi)(z) = \varphi(z) \cdot \left(\overset{l}{\bigwedge}\left(T_z^*\psi\right)\right)\left(\Phi \circ \psi(z)\right)$$

where $T_z^*\psi$ is the adjoint of the tangent map $T\psi$ at z. If $\overline{U}$ is a compact subset of V, the above proposition and corollary apply to the transfer operator $\mathcal{L}$. More generally, these results apply to linear combinations of the form $\int \mu(d\omega)\mathcal{L}_\omega$, under natural conditions on φ_ω, ψ_ω and μ.

We note the following results, which are useful in applications.

LEMMA 12.4. *Let $V \subset \mathbb{C}^d$, V connected and $\psi : V \to U$ be holomorphic, with compact $\overline{U} \subset V$. Then*

$$\bigcap_{l=1}^\infty \psi^k \overline{U}$$

consists of a single point Z. The eigenvalues of the derivative ψ'_Z are strictly less than 1 in modulus.

See [**38**, Lemma 1] (take D bounded open connected, with $D \supset \overline{U}$).

PROPOSITION 12.5. *With the above notation and assumptions*

$$\mathrm{Tr}\,\mathcal{L} = \frac{\varphi(Z)\,\mathrm{tr}\left(\bigwedge^l \psi'_Z\right)}{\det\left(1 - \psi'_Z\right)}$$

see [**38**, Lemma 2].

13. Non-Fredholm situations.

If the dynamical system (M, f) is holomorphic (or real analytic) as well as g, and if f is expanding, the results in the last section can be used to show that $\mathcal{L}$ is analyticity improving. There is therefore a well-defined Fredholm determinant $\mathrm{Det}(1 - z\mathcal{L})$, which is an entire analytic function of z. This leads to a zeta function meromorphic in the whole complex plane.

But there are other situations where $\mathcal{L}$ is not a trace-class operator, or even compact, and one can nevertheless prove that the formal power series $\mathrm{Det}(1 - z\mathcal{L})$ has a non trivial radius of convergence. We shall discuss one example at length in the second part of this monograph, following a general approach which we outline here, and which has proved effective in several different situations.

First one has to make a definite choice for the Banach space B on which $\mathcal{L}$ acts (Hölder, differentiable, or bounded variation functions). One estimates then the *spectral radius* of $\mathcal{L}$ by the formula

$$\text{spectral radius} = \lim_{m \to \infty} \|\mathcal{L}^m\|^{1/m}.$$

Let r be such that there are only finitely many eigenvalues λ of $\mathcal{L}$ (each of finite multiplicity) with $|\lambda| > r$. The infimum of such $r's$ is the *essential spectral radius* and

$$\text{ess. spectral radius} \leq \lim_{m \to \infty} \|\mathcal{L}^m - E_m\|^{1/m}$$

when the E_m are finite rank operators (Nussbaum [31] has shown that for suitable choice of the E_m, the right-hand side is in fact the essential spectral radius). With luck and a clever choice of the E_m one gets an estimate of the essential spectral radius that is strictly less than the spectral radius, and therefore non trivial spectral information.

There is a trick due to N. Haydn [19] which has permitted to show in several cases that $\mathrm{Det}(1 - z\mathcal{L})$ is an analytic function of z for

$$|z| < (\text{ess. spectral radius})^{-1}.$$

Furthermore the zeros of $z \to \mathrm{Det}(1 - z\mathcal{L})$ in the above region are precisely the inverses of the eigenvalues of $\mathcal{L}$, with the same multiplicity. (An explicit example will be discussed in Chapter 2 of this monograph).

Finally, when the weight g is positive, there is "normally" an eigenvalue λ_0 of $\mathcal{L}$ equal to the spectral radius, and which has the expression

$$\lambda = \exp P(\log g)$$

where P is the *pressure* to be described in the next section.

The program oulined above has been made to work in a certain number of examples that we review now briefly.

First let us note that the Fredholm theory (for analyticity improving operators) applies to analytic expanding maps (Ruelle [38], Fried [12]), and to a large class of rational maps of the Riemann sphere (Levin, Sodin and Yuditskii [23, 24]). It can also be made to work (using Markov partitions) for hyperbolic maps and flows when the stable and unstable foliations are analytic (Ruelle [38], Fried [12]). This last condition can however be relaxed as noted by Rugh [45] and Fried.

To study expanding or hyperbolic dynamical systems that are not holomorphic, but only differentiable or Hölder, it is natural to use Markov partitions (introduced by Sinai, Ratner, Bowen). This reduces the study of the original dynamical system to *symbolic dynamics*, i.e., to the study of a subshift of finite type (see Section 2). For this approach we refer to the monograph of Parry and Pollicott [33], and in particular to the references given there to the work of Ruelle, Pollicott, Haydn, etc.

The use of symbolic dynamics has however disadvantages: it is not canonical, and it neglects some of the information contained in the differentiability assumptions. This situation has been progressively improved in a series of papers by Tangerman [47], Ruelle [40, 41], and Fried [14].

Nontrivial analyticity results have also been obtained for the zeta functions associated with piecewise monotone maps of the interval. For such maps, Hofbauer has constructed a "Markov extension" (in effect an infinite Markov partition), and the dynamics has been studied in great detail by Hofbauer and Keller (and many others in different directions). The first result on zeta functions is due to Baladi and Keller [4]. For further work see Keller and Nowicki [22], and Ruelle [43]. Chapter 2 of the present monograph proves an extended version of the Baladi-Keller theorem.

Another approach to zeta functions for piecewise monotone maps of the interval originates with the work of Milnor and Thurston [29]; see also Preston [34], Baladi and Ruelle [5].

14. Thermodynamic formalism.

If ρ is an invariant probability measure for the dynamical system (M, f), the *entropy* (Kolmogorov-Sinai invariant) $h(\rho) = h_f(\rho)$ measures the creation of information by f with respect to ρ (see Billingsley [7]). If M is compact, f continuous, and $A : M \to \mathbb{R}$ is also continuous, an interesting quantity to consider is the *pressure* (see Ruelle [35, 39],

Walters [**48, 49**], Denker, Grillenberger and Sigmund [**11**])

$$P(A) = \sup_{\rho}\big(h(\rho) + \rho(A)\big).$$

In various cases one can prove that h is upper semi continuous (with respect to the vague topology on measures) and the sup is reached by some measures called *equilibrium measures*. (For a general discussion see Bowen [**8**], Ruelle [**39**]).

In Section 13 we have stated the fact that "normally", when $g > 0$, the number $\exp P(\log g)$ is an eigenvalue λ_0 of the transfer operator $\mathcal{L}$, and equal to the spectral radius. In fact "normally", λ_0 is a simple eigenvalue, and more is true: $\mathcal{L}$ has an eigenvector $\Phi > 0$ corresponding to λ_0, and the adjoint $\mathcal{L}^*$ has an eigenvector μ which is a positive measure. Furthermore, (under the normalization condition $\mu(\Phi) = 1$) the product $\Phi \cdot \mu$ is the unique equilibrium state for $\log g$.

We can at least check that $\rho = \Phi \cdot \mu$ is an f-invariant measure, i.e., $\rho(A \circ f) = \rho(A)$ for all continuous $A : M \to \mathbb{R}$. We have indeed

$$\rho(A \circ f) = \mu\big(\Phi \cdot (A \circ f)\big) = \lambda_0^{-1}(\mathcal{L}^*\mu)\big(\Phi \cdot (A \circ f)\big)$$

$$= \lambda_0^{-1}\mu\Big(\mathcal{L}\big(\Phi \cdot (A \circ f)\big)\Big)$$

$$= \lambda_0^{-1}\mu\Big(A \cdot \mathcal{L}(\Phi)\Big)$$

$$= \mu(A \cdot \Phi) = \rho(A)$$

where we have made use of an identity stated at the beginning of Section 8.

An interesting question is that of *decay of correlations* for the equilibrium state $\rho = \Phi \cdot \mu$: does the *correlation function*

$$C(n) = \rho\big(A \cdot (B \circ f^n)\big)$$

decay exponentially for $n \to \infty$? (We take $\rho(A) = \rho(B) = 0$ for simplicity). To study this question we analyze the Fourier-Laplace transform

$$\sum_{n \geq 0} e^{n\alpha} C(n) = \sum_{n \geq 0} e^{n\alpha} \lambda_0^{-n}(\mathcal{L}^{*n}\mu)\big(\Phi A \cdot (B \circ f^n)\big)$$

$$= \sum_{n \geq 0} e^{n\alpha} \lambda_0^{-n} \mu\Big(\mathcal{L}^n\big(\Phi A \cdot (B \circ f^n)\big)\Big)$$

$$= \sum_{n \geq 0} e^{n\alpha} \lambda_0^{-n} \mu\big(B \cdot \mathcal{L}^n(\Phi A)\big)$$

$$= \mu\Big(B\big(1 - e^{\alpha}\lambda_0^{-1}\mathcal{L}\big)^{-1}(\Phi A)\Big).$$

Since the resolvent of $\mathcal{L}$ appears in the right-hand side, we see that the decay of correlations is linked with the spectral properties of the transfer operator, and thus with the analytic properties of the corresponding dynamical zeta function.

15. Ties with other parts of mathematics.

The Riemann zeta function was introduced to study statistical properties of prime numbers. We have seen in the last section that dynamical zeta functions are related to the *thermodynamic formalism*, hence to ergodic theory and again to statistical properties. This gives an idea of the relations that dynamical zeta functions have with more traditional areas of mathematics.

For instance Parry and Pollicott [32] have studied the zeta functions associated with the geodesic flow on a manifold of negative curvature (not necessarily constant) and obtained a theorem on the distribution of closed orbits which is analogous to the prime number theorem.

Another example is Mayer's [27] study of zeta functions associated on one hand with geodesics on the modular surface, on the other with the continued fraction transformation (the equilibrium measure here is the Gauss measure).

Let us conclude with a problem that appears completely open. It is known that the geodesic flow on a compact surface of constant negative curvature is exponentially mixing (Ratner). Does this remain true for non constant negative curvature? (We have seen that the decay of correlations for maps is simply related to spectral properties of a transfer operator, but the situation for flows appears much less tractable).

Piecewise Monotone Maps

In this part we shall study zeta functions associated with systems (X, f, g) where X is a compact subset of $\mathbb{R}$, $f : X \to X$ is piecewise monotone, and $g : X \to \mathbb{C}$ is of bounded variation.

Some references on piecewise monotone maps relevant for the problems discussed here are Hofbauer [20], Hofbauer and Keller [21], and Milnor and Thurston [29]. There is also a vast literature on other aspects of the theory of maps of the interval.

The main result on zeta functions is due to Baladi and Keller [4]. We shall prove an extension this result here using a new method.

An announcement of the results in this Chapter 2 is in Bull. A.M.S. (to appear).

1. Definitions.

Let X be an ordered topological space equivalent, for its order and topology, to a compact subset of $\mathbb{R}$. For simplicity we shall say that X is a *compact subset of* $\mathbb{R}$. An important example is the interval $[0, 1] \subset \mathbb{R}$.

We say that J is an *interval of* X is $J = X \cap I$ where I is an interval of $\mathbb{R}$. If I is closed then J is a *closed interval of* X, i.e., $J = \phi$ or

$$J = \{x \in X : u \leq x \leq v\}$$

for suitable $u, v \in X$, $u \leq v$. A map $f : J \to X$ is *strictly monotone* if it is strictly increasing $\left(x < y \Rightarrow f(x) < f(y)\right)$ or strictly decreasing $\left(x < y \Rightarrow f(x) > f(y)\right)$. If furthermore fJ is an interval of X (i.e., f takes all intermediate values between fu and fv) we say that f has the *Darboux property*; in particular f is continuous and therefore a homeomorphism[1].

We say that $f : X \to X$ is *piecewise monotone* if X is covered by closed intervals $J_1, \ldots, J_N$ such that $f|J_i$ is strictly monotone and

---

[1] A strictly monotone map with the Darboux property is the same thing as a monotone homeomorphism of an interval to an interval.

has the Darboux property for $i = 1, \ldots, N$ (therefore $f|J_i$ is a monotone homeomorphism of J_i to a subinterval of X). We assume that $(J_1, \ldots, J_N)$ is a *minimal cover* of X by closed intervals, i.e., if another such cover $(J'_1, \ldots, J'_N)$ satisfies $J'_1 \subset J_1, \ldots, J'_N \subset J_N$, then $(J'_1, \ldots, J'_N) = (J_1, \ldots, J_N)$. (In particular $J_i \cap J_j$ consists of at most one point when $i \neq j$; of course a partition is a minimal cover). We also assume that no J_i is empty, that $J_1 \leq J_2 \leq \cdots \leq J_N$, and that $N > 1$.

If J_i is reduced to a point we make an arbitrary decision that $f|J_i$ is either increasing, or decreasing (this will be convenient later).

Let $\{b_1, \ldots, b_s\}$ be the set of common endpoints of intervals J_i, J_{i+1}. We define $\varepsilon(x) = 0$ if $x \in \{b_1, \ldots, b_s\}$, and $\varepsilon(x) = \pm 1$ if $x \notin \{b_1, \ldots, b_s\}$ depending on whether f is increasing or decreasing on $J_i \ni x$. The set Per $f = \bigcup_{m \geq 1}$ Fix f^m contains the following subsets:

$$\text{Fix}^\pm f^m = \left\{ x \in \text{Fix } f^m : \prod_{k=0}^{m-1} \varepsilon(f^k x) = \pm 1 \right\}$$

$$\text{Per}^\pm(f, m) = \{ x \in \text{Fix}^\pm f^m : m \text{ is the minimal period of } x \}.$$

We call x a *negative* (resp. *positive*) *periodic point* if $x \in \text{Per}^-(f, m)$ (resp. $x \in \text{Per}^+(f, m)$) for some m. If $(J_1, \ldots, J_N)$ is a partition every periodic point is either negative or positive; in general there may be finitely many exceptions.

We say that $(J_1, \ldots, J_N)$ is a *Markov²* cover for f if, for every i, $f J_i$ is a union of intervals J_j. We say that $(J_1, \ldots, J_N)$ is a *generating cover* if every intersection

$$\bigcap_{n=0}^{\infty} f^{-n} J_{i(n)}$$

consists of at most one point.

Given $g : X \to \mathbb{C}$, where card $X > 1$, we let

$$\text{var } g = \sup \sum_{1}^{n} |g(a_i) - g(a_{i-1})|$$

where the sup is over finite subsets of points of X, ordered such that $a_0 < a_1 < \cdots < a_n$. We say that g is of *bounded variation* if var $g < \infty$. We can similarly define var$(g|\Lambda)$ where Λ is any interval of X (not necessarily closed).

---

²By this definition a Markov partition need not be generating.

The space $\mathcal{B}$ of functions of bounded variation $\Phi : X \to \mathbb{C}$ is a Banach space with respect to the norm Var defined by

$$\text{Var } \Phi = \sup \left(|\Phi(a_0)| + \sum_1^n |\Phi(a_i) - \Phi(a_{i-1})| + |\Phi(a_n)| \right)$$

where the sup is over finite subsets of points of X with $a_0 < a_1 < \cdots < a_n$. Note that in the definition of var and Var the sup may be replaced by a lim over finite sets ordered by inclusion.

If $X = \{a\}$, we let $\text{var } g = 0$ and $\text{Var } \Phi = 2|\Phi(a)|$. With these definitions the norm Var is equivalent to the norm $\| \cdot \|_0 + \text{var}$. In fact

$$\|\Phi\|_0 \le \frac{1}{2} \text{Var } \Phi$$
$$\text{var } \Phi \le \text{Var } \Phi$$
$$\text{Var } \Phi \le 2\|\Phi\|_0 + \text{var } \Phi.$$

We have also

$$\text{Var}(\Phi \cdot \Psi) \le \|\Phi\|_0 \text{Var } \Psi + \text{var } \Phi \|\Psi\|_0 \le \text{Var } \Phi \cdot \text{Var } \Psi$$

and

$$\text{Var}(\Phi \circ \psi) = \text{Var } \Phi$$

if ψ is strictly monotone and has the Darboux property.

Given $Y \subset X$ we shall denote by $\mathcal{B}_Y$ the subspace of $\mathcal{B}$ consisting of the functions Φ vanishing outside Y, and by $\mathcal{B}_{\backslash Y}$ the quotient Banach space $\mathcal{B}/\mathcal{B}_Y$.

2. Construction of new systems.

Let the system (X, f, g) consist of a compact subset X of $\mathbb{R}$, a piecewise monotone map f with an associated minimal cover by closed intervals $(J_1, \ldots, J_N)$, and a function g of bounded variation. From these data we shall construct in various ways a new system $(\widehat{X}, \hat{f}, \hat{g})$, and a cover $(\hat{J}_1, \ldots, \hat{J}_N)$ with desirable properties: the intervals $(\hat{J}_1, \ldots, \hat{J}_N)$ form a partition, or $(\hat{J}_1, \ldots, \hat{J}_N)$ is Markov, or generating, or $\hat{g}$ is continuous at periodic points.

The set $\widehat{X}$ obtained by construction is in general not an interval of $\mathbb{R}$ (even if X is); this explains why we do not want to restrict attention to such intervals.

Remember that $\mathcal{B}$ denotes the Banach space of functions of bounded variation $X \to \mathbb{C}$. We denote by $\widehat{\mathcal{B}}$ the similar space of functions of bounded variation $\widehat{X} \to \mathbb{C}$. We also let $\widehat{\mathcal{B}}_{\widehat{Y}}$ be the subspace of functions vanishing outside of a set $\widehat{Y} \subset \widehat{X}$, and use the notation $\widehat{\mathcal{B}}_{\backslash \widehat{Y}} = \widehat{\mathcal{B}}/\widehat{\mathcal{B}}_{\widehat{Y}}$.

PROPOSITION 2.1 (Producing a partition $(\hat{J}_1, \ldots, \hat{J}_N)^3$). *We may choose* $(\widehat{X}, \hat{f}, \hat{g})$, $(\hat{J}_1, \ldots, \hat{J}_N)$, *and* $\hat{\pi} : \widehat{X} \to X$ *order-preserving continuous surjective such that* $\hat{\pi} \circ \hat{f} = f \circ \hat{\pi}$, $\hat{g} = g \circ \hat{\pi}$, *and* $\hat{\pi}\hat{J}_i = J_i$ *for* $i = 1, \ldots, N$. *Furthermore* $\hat{J}_1, \ldots, \hat{J}_N$ *are disjoint and* $\hat{\pi}$ *is two-to-one on a countable set, one-to-one elsewhere.*

If $Z = \{x \in X, \operatorname{card} \hat{\pi}^{-1}x = 2\}$, $Y = \bigcup_{n \geq 0} f^n Z$, *and* $\widehat{Y} = \hat{\pi}^{-1}Y$, *each point of* Y *is a limit of points in* $X \backslash Y$ *and each point of* $\widehat{Y}$ *is a limit of points in* $\widehat{X} \backslash \widehat{Y}$.

The map $\Phi \mapsto \Phi \circ \hat{\pi}$ *defines an isomorphism* $\mathcal{B}_{\backslash Y} \to \widehat{\mathcal{B}}_{\backslash \widehat{Y}}$ *of Banach spaces.*

Let $b_1, \ldots, b_s$ be the points of X belonging to two distinct intervals J_i. Suppose that $\xi \in X$, that $f^k \xi \in \{b_1, \ldots, b_s\}$ for some $k \geq 0$, and choose the smallest such k. If $k \geq 1$ also assume that ξ is not an endpoint of one of the J_i. Under these conditions replace ξ by two points $\xi_- < \xi_+$ and insert a gap of length $\varepsilon \alpha^k$ between them. We assume $0 < \alpha < 1/N$ so that the total length of the inserted gaps is $\leq s\varepsilon(1 - N\alpha)^{-1}$. We obtain thus a compact set $\widehat{X} \subset \mathbb{R}$; the collapse map $\hat{\pi} : \widehat{X} \to X$ is order preserving, and $\hat{\pi}$ is two-to-one on a countable set, one-to-one elsewhere.

Defining $\hat{g}$ by $\hat{g} = g \circ \hat{\pi}$ we see that $\operatorname{var} \hat{g} = \operatorname{var} g < \infty$.

If $J_i = \{x \in X : \alpha_i \leq x \leq \beta_i\}$ we define $\hat{J}_i = \{\xi \in \widehat{X} : \hat{\alpha}_i \leq \xi \leq \hat{\beta}_i\}$ where $\hat{\pi}\hat{\alpha}_i = \alpha_i$, $\hat{\pi}\hat{\beta}_i = \beta_i$; if $\hat{\pi}^{-1}\alpha_i = \{\alpha_{i-}, \alpha_{i+}\}$ we take $\hat{\alpha}_i = \alpha_{i+}$; if $\hat{\pi}^{-1}\beta_i = \{\beta_{i-}, \beta_{i+}\}$ we take $\hat{\beta}_i = \beta_{i-}$. With this definition the $\hat{J}_i$ are disjoint and $\hat{\pi}\hat{J}_i = J_i$.

The map $\hat{f}$ is entirely specified by the condition that $\hat{f}$ be a monotone homeomorphism of $\hat{J}_i$ to a closed interval of $\widehat{X}$ for $i = 1, \ldots, N$, and that $\hat{\pi} \circ \hat{f} = f \circ \hat{\pi}$.

We will now show that points of Y (resp. $\widehat{Y}$) are limits of points of $X \backslash Y$ (resp. $\widehat{X} \backslash \widehat{Y}$). Since $(J_1, \ldots, J_N)$ is a minimal cover, the points $b_1, \ldots, b_s$ belonging to two different J_i must be limit points from the left and from the right of other points of X. Remember that

$$Y = \bigcup_{n \geq 0} f^n Z$$

and

$$Z = \left\{ x \in X : \operatorname{card} \hat{\pi}^{-1}x = 2 \right\} = \bigcup_{k \geq 0} Z_k$$

where

$$Z_0 = \{b_1, \ldots, b_s\}$$

---

[3]See Hofbauer and Keller [21] for the case $X = [0, 1]$.

and, for $k \geq 1$,

$$Z_k = f^{-1}Z_{k-1} \backslash Z_k^*$$
$$Z_k^* = \{\text{endpoints of intervals } J_i\} \cup Z_1 \cup \cdots \cup Z_{k-1}.$$

By induction on k all points of Z are limit points (from the left and from the right) of other points of X. Therefore all points of Y are limits (from one side at least) of other points of X.

Suppose now that $y \in Y$ and y is not a limit of points in $X \backslash Y$, i.e., there is an open neighborhood V of y such that $X \cap V = Y \cap V$. Every points in $Y \cap V$ is thus a limit of other points in $Y \cap V$. For all $n \geq 0$ we can therefore construct (by successive doubling) finite sets $Y_n, Y_n' \subset Y$ such that $\operatorname{card} Y_n = \operatorname{card} Y_n' = 2^n$, and $Y_0 = \{y\}$, $Y_n' \cap Y_n = \phi$, $Y_{n+1} = Y_n \cup Y_n'$. We may assume that for each n there is $\varepsilon_n > 0$ such that the mutual distance of points of Y_n is $> \varepsilon_n$ and each point of Y_n' has distance $\leq \frac{1}{3}\varepsilon_n$ to a different point of Y_n. The closure of $\bigcup_{n \geq 0} Y_n$ is then a Cantor set $K \subset X$ and $K \cap V$ is uncountable contrary to the assumption that $K \cap V \subset X \cap V = Y \cap V$, which is countable. Therefore Y is in the closure of $X \backslash Y$.

To obtain the corresponding result for $\widehat{Y} = \hat{\pi}^{-1}Y$, notice that $\widehat{Y} = \bigcup_{n \geq 0} \hat{f}^n \widehat{Z}$ where $\widehat{Z} = \hat{\pi}^{-1}Z$ and all points of $\widehat{Z}$ are limits (from one side) of other points of $\widehat{X}$. Therefore all points of $\widehat{Y}$ are limits (from one side at least) of other points of $\widehat{X}$, and the proof proceeds as before.

The map $\Phi \mapsto \Phi \circ \hat{\pi}$ identifies $\mathcal{B}$ with the subspace of $\widehat{\mathcal{B}}$ consisting of those $\widehat{\Phi}$ such that $\widehat{\Phi}(\xi) = \widehat{\Phi}(\xi')$ when $\hat{\pi}\xi = \hat{\pi}\xi'$. Given $\widehat{\Psi} \in \widehat{\mathcal{B}}_{\backslash \widehat{Y}}$ let $\widehat{\Phi} \in \widehat{\mathcal{B}}$ be in the class of $\widehat{\Psi}$; we may change $\widehat{\Phi}$ at the points of $\widehat{Y}$ to obtain $\widehat{\Phi}'$ such that $\widehat{\Phi}'(\xi) = \widehat{\Phi}'(\xi')$ when $\hat{\pi}\xi = \hat{\pi}\xi'$, and $\operatorname{Var} \widehat{\Phi}' \leq \operatorname{Var} \widehat{\Phi}$. Therefore $\widehat{\Phi}' = \Phi \circ \hat{\pi}$ and, if $\Psi \in \mathcal{B}_{\backslash Y}$ is the class of Φ,

$$\|\widehat{\Psi}\| = \inf \operatorname{Var} \widehat{\Phi}' = \inf \operatorname{Var} \Phi \circ \hat{\pi}$$
$$= \inf \operatorname{Var} \Phi = \|\Psi\|.$$

The map $\Psi \mapsto \widehat{\Psi}$ is thus an isomorphism $\mathcal{B}_{\backslash Y} \to \widehat{\mathcal{B}}_{\backslash \widehat{Y}}$ of Banach spaces.

Remarks 2.2. (1) Except for the finitely many periodic orbits through $b_1, \ldots, b_s$, the map $\hat{\pi}$ defines a bijection $\operatorname{Fix} \hat{f}^m \to \operatorname{Fix} f^m$ for every integer $m \geq 1$.

(2) Since the points $b_1, \ldots, b_s$ are doubled by the construction, it is possible to start with "functions" f and g which are two-valued at these points (with a "left value" and a "right value"). In this manner one can deal with piecewise monotone maps $f : [0,1] \to [0,1]$ with discontinuities.

PROPOSITION 2.3 (Producing a Markov partition $(\hat{J}_1, \ldots, \hat{J}_N)^4$).
If $(J_1, \ldots, J_N)$ is a partition, we may choose $(\widehat{X}, \hat{f}, \hat{g})$, $(\hat{J}_1, \ldots, \hat{J}_N)$, and $\pi : X \to \widehat{X}$ order-preserving continuous injective such that $\hat{f} \circ \pi = \pi \circ f$, $\hat{g} \circ \pi = g$, and $\pi J_i \subset \hat{J}_i$ for $i = 1, \ldots, N$. Furthermore $(\hat{J}_1, \ldots, \hat{J}_N)$ is a Markov partition.

Use now π to identify X with a subset of $\widehat{X}$ so that $\hat{f}$, $\hat{g}$ extend f, g and $J_i = \hat{J}_i \cap X$ for $i = 1, \ldots, N$. Then $\widehat{Y} = \widehat{X} \backslash X$ is the union of a family (U_α) of disjoint open intervals, each contained in some $\hat{J}_i$, and for each U_α there is $n \geq 0$ such that U_α, $\hat{f} U_\alpha$, $\ldots$, $\hat{f}^n U_\alpha$ are intervals in the family (U_α) and $\hat{g}|\hat{f}^n U_\alpha = 0$. Each interval U_α is separated either from the U_β below it or from those above it by a point $x \in X = \widehat{X} \backslash \widehat{Y}$.

The map $\widehat{\Phi} \mapsto \widehat{\Phi}|X$ defines an isomorphism $\widehat{\mathcal{B}}_{\backslash \widehat{Y}} \to \mathcal{B}$ of Banach spaces.

We recall the assumption $J_1 < \cdots < J_N$ and write $\varepsilon(i) = \pm 1$ depending on whether f is increasing or decreasing on J_i.

Let $X_{i_1 \ldots i_k}$, $J_{i_1 \ldots i_k i}$ be copies of X and J_i with the original order if $\prod_{r=1}^{k} \varepsilon(i_r) = +1$ and the reverse order if $\prod_{r=1}^{k} \varepsilon(i_r) = -1$. Supposing $i_1 = j_1, \ldots, i_{l-1} = j_{l-1}$ and $i_l < j_l$, let

$$X_{i_1 \ldots i_k} < X_{j_1 \ldots j_k} \quad \text{if} \quad \prod_{r=1}^{l-1} \varepsilon(i_r) = +1$$

$$X_{i_1 \ldots i_k} > X_{j_1 \ldots j_k} \quad \text{if} \quad \prod_{r=1}^{l-1} \varepsilon(i_r) = -1.$$

This defines an order on the disjoint union

$$X^{(k)} = \bigcup_{i_1 \ldots i_k} X_{i_1 \ldots i_k} = \bigcup_{i_1 \ldots i_k i_{k+1}} J_{i_1 \ldots i_k i_{k+1}}$$

for $k \geq 1$; we also write $X^{(0)} = X$.

The restriction of f to J_i defines a map $J_i \to X$, hence also

$$J_{i_1 \ldots i_k i_{k+1}} \to X_{i_1 \ldots i_{k+1}}$$

and therefore also a map

$$\pi^{(k)} : X^{(k)} = \bigcup J_{i_1 \ldots i_k i_{k+1}} \to X^{(k+1)}$$

which is an order preserving homeomorphism of $X^{(k)}$ to its image $\pi^{(k)} X^{(k)} \subset X^{(k+1)}$.

---

[4]See the Appendix of Ruelle [43] for the case where $(J_1, \ldots, J_N)$ is generating.

Let $\hat{\pi}^{(k)} : X^{(k+1)} \to X^{(k)}$ be defined by

$$\hat{\pi}^{(k)}(x) = \begin{cases} \left(\pi^{(k)}\right)^{-1} x & \text{if} \quad x \in \pi^{(k)} J_{i_1 \dots i_k i_{k+1}} \\ \min \text{ or } \max J_{i_1 \dots i_k i_{k+1}} & \text{if} \quad x \in X_{i_1 \dots i_{k+1}} \backslash \pi^{(k)} J_{i_1 \dots i_k i_{k+1}} \end{cases}$$

where the choice of min or max is made so that $\hat{\pi}^{(k)}$ is non decreasing. In particular

$$\hat{\pi}^{(k)} \pi^{(k)} = \text{ identity of } X^{(k)}.$$

We define $\widehat{X} = \varprojlim X^{(k)}$ to be the inverse limit of

$$X^{(0)} \xleftarrow{\hat{\pi}^{(0)}} X^{(1)} \longleftarrow \cdots \longleftarrow X^{(k)} \xleftarrow{\hat{\pi}^{(k)}} X^{(k+1)} \longleftarrow \cdots$$

and let $\hat{\pi} : \widehat{X} \to X^{(0)} = X$ be the associated map. The set $\widehat{X}$ is compact, ordered, and may be viewed as a compact subset of $\mathbb{R}$; $\hat{\pi}$ is continuous and order preserving. We also define $\pi : X \to \widehat{X}$ by

$$\pi x = \left(x, \pi^{(0)} x, \pi^{(1)} \pi^{(0)} x, \dots \right).$$

This is an order preserving homeomorphism of X to its image, and

$$\hat{\pi} \pi = \text{ identity of } X.$$

Write $J_i^{(0)} = J_i$ and

$$J_i^{(k+1)} = \bigcup_{i_1 \dots i_k} X_{i i_1 \dots i_k} \subset X^{(k+1)}$$

for $k + 1 \geq 1$. Since $X_{i i_1 \dots i_k}$ and $X_{i_1 \dots i_k}$ are copies of X there is a naturally defined monotone homeomorphism of $J_i^{(k+1)}$ onto $X^{(k)}$, hence a piecewise monotone map

$$f^{(k)} : X^{(k+1)} = \bigcup_i J_i^{(k+1)} \to X^{(k)}$$

and it is easily verified that

$$f^{(k+1)} \circ \pi^{(k+1)} = \pi^{(k)} \circ f^{(k)}$$
$$f^{(k)} \circ \hat{\pi}^{(k+1)} = \hat{\pi}^{(k)} \circ f^{(k+1)}.$$

If $\xi = \{x_0, x_1, \dots, x_n, \dots) \in \widehat{X}$, let

$$\hat{f}\xi = \left(f^{(0)} x_1, f^{(1)} x_2, \dots, f^{(n)} x_{n+1}, \dots \right).$$

The set

$$\hat{J}_i = \varprojlim J_i^{(k)} = \hat{\pi}^{-1} J_i$$

is mapped by $\hat{f}$ homeomorphically onto $\widehat{X}$, preserving or reversing the order according to whether $\varepsilon(i) = \pm 1$; $\hat{f}$ is thus piecewise monotone $\widehat{X} \to \widehat{X}$, we have

$$\hat{f} \circ \pi = \pi \circ f$$

and $(\hat{J}_1, \ldots, \hat{J}_N)$ is a Markov partition for $\hat{f}$. Furthermore $\hat{J}_i = \hat{\pi}^{-1}(\hat{\pi} \circ \pi) J_i \supset \pi J_i$.

Let χ_1 be the characteristic function of the union of the N intervals $[\min \pi J_i, \max \pi J_i]$ in $\widehat{X}$. We define then

$$\hat{g}(\xi) = \chi_1(\xi) \circ g(\hat{\pi}\xi)$$

which is of bounded variation (because $\operatorname{Var}\chi_1 \le 2N$ and $\operatorname{Var} g \circ \hat{\pi} = \operatorname{Var} g$).

If α is any triple (n, i, ε) where $n \ge 0$, $1 \le i \le N$, and $\varepsilon = \pm 1$ we let

$$U_\alpha = \Big\{ \xi = (x_0, x_1, \ldots) \in \hat{J}_i :$$
$$x_{k+1} = \pi^{(k)} x_k \text{ for } k = 0, \ldots, n-1 \text{ and } x_{n+1} \lesssim \pi^{(n)} x_n \Big\}$$

where $\lesssim$ means $<$ if $\varepsilon = -1$, $>$ if $\varepsilon = +1$. We have thus $\bigcup_\alpha U_\alpha = \widehat{X} \backslash \hat{\pi} X = \widehat{Y}$. If $n > 0$, fU_α is an interval U_β in the family (U_α). [To see this use $f^{(k+1)} \pi^{(k+1)} = \pi^{(k)} f^{(k)}$]. If $n = 0$, then $\xi \in U_\alpha$ means $x_1 \notin \pi^{(0)} J_i$, i.e., $\chi_1(\xi) = 0$ hence $\hat{g}(\xi) = 0$.

Note also that U_α is separated from the U_β either above it or below it by the point $\pi\hat{\pi} U_\alpha$.

The restriction $\widehat{\Phi} \mapsto \widehat{\Phi} \circ \pi$ defines a norm reducing map $\widehat{\mathcal{B}} \to \mathcal{B}$, but for each $\Phi \in \mathcal{B}$ one can find $\widehat{\Phi}$ such that $\operatorname{Var}\widehat{\Phi} = \operatorname{Var}\Phi$ and $\widehat{\Phi} \circ \pi = \Phi$, therefore $\widehat{\Phi} \mapsto \widehat{\Phi} \circ \pi$ defines an isomorphism $\widehat{\mathcal{B}}_{\backslash \widehat{Y}} \to \mathcal{B}$.

PROPOSITION 2.4 (Producing a generating partition $(\hat{J}_1, \ldots, \hat{J}_N)$[5]). *If $(J_1, \ldots, J_N)$ is a partition, we may choose $(\widehat{X}, \hat{f}, \hat{g})$, $(\hat{J}_1, \ldots, \hat{J}_N)$, and $\pi : X \to \widehat{X}$ order preserving continuous surjective such that $\hat{f} \circ \pi = \pi \circ f$, $\hat{g}(\xi) = g(\pi^{-1}\xi)$ when* $\operatorname{card}\pi^{-1}\xi = 1$[6]*, and $\pi J_i = \hat{J}_i$ for $i = 1, \ldots, N$. Furthermore $(\hat{J}_1, \ldots, \hat{J}_N)$ is a generating partition.*

The set $\widehat{Y} = \{\xi \in \widehat{X} : \operatorname{card}\pi^{-1}\xi > 1\}$ is countable and the closed intervals $U_\alpha = \pi^{-1}\xi$ with $\xi \in \widehat{Y}$ are of the form $\bigcap_{k \ge 0} f^{-k} J_{i(k)}$. Each U_α is mapped by f into another interval U_β of the same family: $fU_\alpha \subset U_\beta$. If we write $Y = \pi^{-1}\widehat{Y} = \bigcup_\alpha U_\alpha$, the map $\widehat{\Phi} \to \widehat{\Phi} \circ \pi$ defines an isomorphism $\widehat{\mathcal{B}}_{\backslash \widehat{Y}} \to \mathcal{B}_{\backslash Y}$ of Banach spaces.

---

[5] See Baladi and Ruelle [**6**] for the case of intervals of $\mathbb{R}$.

[6] Further conditions will be imposed in Remark 2.5 (1) and Proposition (3.2) below.

If $(J_1, \ldots, J_N)$ is a Markov partition, then $(\hat{J}_1, \ldots, \hat{J}_N)$ is a generating Markov partition. Each $\xi \in \hat{Y}$ is a limit of points in $\widehat{X} \backslash \hat{Y}$ and, for each interval U_α, either the upper or the lower endpoint is a limit of points in $X \backslash Y$.

For $x, y \in X$ we write $x \sim y$ if $f^k x, f^k y$ belong to the same $J_{i(k)}$ for all $k \geq 0$. This is an equivalence relation because we have assumed that $J_1, \ldots, J_N$ are disjoint. Every equivalence class $[x]$ is a closed interval which we call U_α if it contains more than one point. We have thus diam $U_\alpha > 0$, which can happen only for countably many U_α. Let $\pi : X \to \widehat{X}$ collapse each U_α to a point; $\widehat{X}$ is thus a compact subset of $\mathbb{R}$, π is order preserving, and card $\pi^{-1} x = 1$ outside of a countable set $\hat{Y}$.

The properties $\hat{f} \circ \pi = \pi \circ f$ and $\pi J_i = \hat{J}_i$ define $\hat{f}$ and a partition $(\hat{J}_1, \ldots, \hat{J}_N)$ associated with $\hat{f}$. By construction, this partition is generating.

For the moment we only impose on $\hat{g}$ the condition $\hat{g}(\xi) = g(\pi^{-1}\xi)$ when card $\pi^{-1}\xi = 1$, this is certainly compatible with Var $\hat{g} < \infty$.

Because f restricted to $U_\alpha (\subset J_i)$ is a homeomorphism, f maps $U_\alpha = \bigcap_{k \geq 0} f^{-k} J_{i(k)}$ into $\bigcap_{k \geq 1} f^{-k+1} J_{i(k)} = \bigcap_{k \geq 0} f^{-k} J_{i(k+1)} = U_\beta$, where card $U_\beta > 1$.

The map $\hat{\Phi} \to \hat{\Phi} \circ \pi$ defines a norm reducing map $\hat{\mathcal{B}}_{\backslash \hat{Y}} \to \mathcal{B}_{\backslash Y}$. But if $\Psi \in \mathcal{B}_{\backslash Y}$ there is a Φ in the class of Ψ which is constant on each U_α and such that Var $\Phi = \|\Psi\|$; we may thus write $\Phi = \hat{\Phi} \circ \pi$ with Var $\hat{\Phi} = $ Var $\Phi = \|\Psi\|$. This shows that $\hat{\mathcal{B}}_{\backslash \hat{Y}} \to \mathcal{B}_{\backslash Y}$ is isometric.

If $(J_1, \ldots, J_N)$ is a Markov partition, clearly $(\hat{J}_1, \ldots, \hat{J}_N)$ is a Markov partition and $\widehat{X} = \pi X$ is a Cantor set. Since $\hat{Y}$ is countable, every point of $\hat{Y}$ is a limit of points in $\widehat{X} \backslash \hat{Y}$. Therefore also each U_α contains a point which is a limit of points in $X \backslash Y$.

Remarks 2.5. (1) π defines an injective map

$$\mathrm{Fix}^- f^m \to \mathrm{Fix}^- \hat{f}^m$$

and we may take $\hat{g}(\pi x) = g(x)$ for $x \in \mathrm{Fix}^- f^m$.

[Since $\hat{f}^m \circ \pi = \pi \circ f^m$ we have $\pi \, \mathrm{Fix} \, f^m \subset \mathrm{Fix} \, \hat{f}^m$, and also $\pi \, \mathrm{Fix}^- f^m \subset \mathrm{Fix}^- \hat{f}^m$. Suppose that $x \in \mathrm{Fix}^- f^m$, $y \in \mathrm{Fix}^- f^n$ and $\pi x = \pi y = \xi$. We have then $\xi \in \mathrm{Fix}^- \hat{f}^k$ and we may assume that k is the minimal period of ξ, so that $m = kp$, $n = kq$, and p, q are odd. Since f^{kpq} induces a decreasing map $\pi^{-1}\xi \to \pi^{-1}\xi$, the fixed points x and y coincide. Therefore $\pi^{-1}\xi$ contains at most one negative periodic point for f, and the map $\mathrm{Fix}^- f^m \to \mathrm{Fix}^- \hat{f}^m$ is injective. The injectivity of

$\mathrm{Fix}^- f^m \to \mathrm{Fix}^- \hat{f}^m \hookrightarrow \widehat{X}$ allows the definition $\hat{g}(\pi x) = g(x)$ when $x \in \mathrm{Fix}^- f^m$].

(2) If f is piecewise expanding (i.e., if $f|J_i$ expands by a factor $\geq \theta^{-1} > 1$ for $i = 1, \ldots, N$) the partition $(J_1, \ldots, J_N)$ is generating. The map π constructed in the proposition is then the identity.

(3) The constructions of Propositions 2.3 and 2.4 can be applied successively, and they *almost* commute. Independently of order of application one obtains a full shift $(\widehat{X}, \hat{f})$ and a map $\pi : X \to \widehat{X}$ such that $\hat{f} \circ \pi = \pi \circ f$. Depending on the order of application of Propositions 2.3 and 2.4, g is replaced by $\hat{g}_1$ or $\hat{g}_2$ which may be different. But if corresponding choices are made in the construction of $\hat{g}_1$, $\hat{g}_2$, then $\hat{g}_2 - \hat{g}_1$ vanishes on πX.

COROLLARY 2.6. *Let* $0 = a_0 < a_1 < \cdots < a_N = 1$. *We assume that* $f : [0, 1] \to [0, 1]$ *is continuous and strictly monotone on the intervals* $[a_{i-1}, a_i]$, *and that* g *is of bounded variation*[7]. *Let then* $(\widetilde{X}, \tilde{f}, \tilde{g})$, $(\tilde{J}_1, \ldots, \tilde{J}_N)$, $\tilde{\pi}_1$ *be obtained by application of Proposition 2.1 to* $([0, 1]$, f, $g)$, $([a_0, a_1], \ldots, [a_{N-1}, a_N])$; *also let* $(\widehat{X}, \hat{f}, \hat{g})$, $(\hat{J}_1, \ldots, \hat{J}_N)$, $\tilde{\pi}_2$ *be obtained by application of Proposition 2.4 to* $(\widetilde{X}, \tilde{f}, \tilde{g})$, $(\tilde{J}_1, \ldots, \tilde{J}_N)$. *With this notation,* $\tilde{\pi}_2$ *defines a bijection* $\mathrm{Fix}^- \tilde{f}^m \to \mathrm{Fix}^- \hat{f}^m$, *and we may take* $\hat{g}(\tilde{\pi}_2 x) = g(x)$ *for* $x \in \mathrm{Fix}^- \tilde{f}^m$.

By construction each set $\tilde{\pi}_2^{-1} \xi$ is an interval of $\mathbb{R}$ (i.e., connected). If $\xi \in \mathrm{Fix}^- \hat{f}^m$, then $\tilde{f}^m$ maps the interval $\tilde{\pi}_2^{-1} \xi$ into itself, and therefore $\tilde{\pi}_2^{-1} \xi$ contains a fixed point $x \in \mathrm{Fix}^- \tilde{f}^m$. This shows that $\tilde{\pi}_2$ maps $\mathrm{Fix}^- \tilde{f}^m$ onto $\mathrm{Fix}^- \hat{f}^m$, and the corollary follows from Remark 2.5 (1).

PROPOSITION 2.7 (Producing $\hat{g}$ continuous at periodic points). *If* $(J_1, \ldots, J_N)$ *is a generating partition and* S *denotes the set of periodic points, we may choose* $(\widehat{X}, \hat{f}, \hat{g})$, *a partition* $(\hat{J}_1, \ldots, \hat{J}_N)$, *and* $\hat{\pi} : \widehat{X} \to X$ *order preserving continuous surjective such that* $\hat{\pi} \circ \hat{f} = f \circ \hat{\pi}$, $\hat{g}(\xi) = g(\hat{\pi}\xi)$ *if* $\hat{\pi}\xi \notin S$ *and* $\hat{g}$ *is continuous at the points of* $\hat{\pi}^{-1}S$; $\hat{J}_i = \hat{\pi}^{-1}J_i$ *for* $i = 1, \ldots, N$. *The partition* $(\hat{J}_1, \ldots, \hat{J}_N)$ *is in general not generating. The map* $\hat{\pi}$ *is two-to-one on a countable set, one-to-one elsewhere. Furthermore the entropy of* $\hat{f}$-*invariant probability measures is* $\hat{h} = h \circ \pi$, *so that* $\hat{h}$ *is u.s.c.*

There is a countable set Y, *containing* $\{x \in X : \mathrm{card}\, \hat{\pi}^{-1}x = 2\}$ *and the periodic points of discontinuity of* g, *such that* $fY \subset Y$ *and each*

---

[7] f and g may be two-valued at $a_1, \ldots, a_{N-1}$, see Remark 2.2 (2).

point of Y is a limit of points in $X \backslash Y$. Correspondingly, if $\widehat{Y} = \hat{\pi}^{-1}Y$, then $\hat{f}\widehat{Y} \subset \widehat{Y}$ and each point of $\widehat{Y}$ is a limit of points in $\widehat{X} \backslash \widehat{Y}$.

The map $\Phi \mapsto \Phi \circ \hat{\pi}$ defines an isomorphism $\mathcal{B}_{\backslash Y} \to \widehat{\mathcal{B}}_{\backslash \widehat{Y}}$ of Banach spaces.

If X is a Cantor set, then also $\widehat{X}$ is a Cantor set.

Let X_0, X_1, X_2 consist of those points of X which are isolated (X_0), or limits of other points of X on one side only (X_1), or limits of other points of X on both sides (X_2). A point $x \in X_2 \cap J_i$ cannot be an endpoint of J_i, and since $f|J_i$ is a monotone homeomorphism, we have again $fx \in X_2$. Thus $fX_2 \subset X_2$. Similarly $fX_1 \subset X_1 \cup X_2$, and of course $fX_0 \subset X_0 \cup X_1 \cup X_2$.

If we write $S_0 = S \cap X_0$, $S_1 = S \cap X_1$, $S_2 = S \cap X_2$, it follows that $fS_0 = S_0$, $fS_1 = S_1$, $fS_2 = S_2$.

We define $Z_0 = S_2$ and, by induction,

$$Z_k = f^{-1}Z_{k-1} \backslash \left(\{\text{endpoints of intervals } J_i\} \cup Z_0 \right).$$

To construct $\widehat{X}$ we replace each $\xi \in Z = \bigcup_{k \geq 0} Z_k$ by two points $\xi_- < \xi_+$. If $\eta = f^k \xi$, where $\xi \in Z_k$ and η has period l, we insert a gap of length $\varepsilon \alpha^{k+l}$ between ξ_- and ξ_+ (with $0 < \alpha < 1/N$). We let $\hat{\pi}$ be the collapse map $\widehat{X} \to X$ and define $\hat{J}_i = \hat{\pi}^{-1}J_i$ for $i = 1, \ldots, N$. The map $\hat{f}$ is uniquely defined by $\hat{\pi} \circ \hat{f} = f \circ \hat{\pi}$ and the condition that $\hat{f}|\hat{J}_i$ is a monotone homeomorphism of $\hat{J}_i$ to an interval of $\widehat{X}$.

We choose $Y = S_1 \cup Z$. Since g is continuous at the points of S_0, we see that Y contains the periodic points of discontinuity for g; Y also contains $\{x \in X : \text{card } \hat{\pi}^{-1}x = 2\} = Z$. Since $fZ \subset Z$, we have $fY \subset Y$. If we define $\widehat{Y} = \hat{\pi}^{-1}Y$, then $\hat{\pi} \circ \hat{f} = f \circ \hat{\pi}$ yields $\hat{f}\widehat{Y} \subset \widehat{Y}$.

Let us now show that each point in $\widehat{Y}$ is a limit of points in $\widehat{X} \backslash \widehat{Y}$. By assumption each $\xi \in \hat{\pi}^{-1}S_1$, $\hat{\pi}^{-1}S_2$ is a limit (on one side) of other points of $\widehat{X}$. By induction on k this is also true for points $\xi \in \hat{\pi}^{-1}Z_k$ (remember that $Z_0 = S_2$). Therefore each $\xi \in \widehat{Y} = \hat{\pi}^{-1}(S_1 \cup Z)$ is a limit of other points of $\widehat{X}$. As in the proof of Proposition 2.1 this implies that ξ is a limit of points in $\widehat{X} \backslash \widehat{Y}$. [Otherwise there would be a neighborhood V of ξ such that $V \cap \widehat{X} = V \cap \widehat{Y}$ and one could construct a Cantor set $K \subset \widehat{X}$ with $\xi \in K \subset V$ in contradiction with the fact that $\widehat{Y}$ is countable]. This result also implies that each point in Y is a limit of points in $X \backslash Y$.

If $\hat{\pi}\xi \notin S_1 \cup S_2$ we let $\hat{g}(\xi) = g(\hat{\pi}\xi)$. $\hat{\pi}\xi \in S_1 \cup S_2$ then ξ is a one-side limit of points η of $\widehat{X}$ (or even of $\widehat{X} \backslash \widehat{Y}$) and we take $\hat{g}(\xi) = \lim g(\hat{\pi}\eta)$. This limit exists because g has bounded variation. With this definition,

$\hat{g}$ is of bounded variation, continuous at the points of $\hat{\pi}^{-1}S$ and satisfies $\hat{g}(\xi) = g(\hat{\pi}\xi)$ if $\hat{\pi}\xi \notin S$.

If $\hat{\rho}$ is an $\hat{f}$-invariant probability measure we may write $\hat{\rho} = \alpha\hat{\rho}_0 + (1-\alpha)\hat{\rho}_1$ where $\hat{\rho}_0$ (resp. $\hat{\rho}_1$) is an atomic (resp. nonatomic) probability measure. Since the entropy of atomic measures vanishes, and $\hat{\pi}$ is an isomorphism with respect to nonatomic measures, we have

$$\hat{h}(\hat{\rho}) = \alpha\hat{h}(\hat{\rho}_0) + (1 - \alpha)\hat{h}(\hat{\rho}_1)$$
$$= (1 - \alpha)\hat{h}(\hat{\rho}_1) = (1 - \alpha)h(\hat{\pi}\hat{\rho}_1)$$
$$= \alpha h(\hat{\pi}\hat{\rho}_0) + (1 - \alpha)h(\hat{\pi}\hat{\rho}_1)$$
$$= h(\hat{\pi}\hat{\rho}).$$

Since $(J_1, \ldots, J_N)$ is a generating partition, h is u.s.c. for the vague topology[8], and since $\hat{\pi}$ is (vaguely) continuous we see that also $\hat{h}$ is u.s.c.

The proof that $\Phi \mapsto \Phi \circ \hat{\pi}$ defines an isomorphism $\mathcal{B}_{\backslash Y} \to \widehat{\mathcal{B}}_{\backslash \widehat{Y}}$ is easy, and essentially the same as in Proposition 2.1.

Finally, since the construction of $\widehat{X}$ does not create isolated points, $\widehat{X}$ is a Cantor set when X is.

3. The functional Θ.

Given a compact subset X of $\mathbb{R}$, and a piecewise monotone map f, we shall define a functional $g \mapsto \Theta$ on functions g of bounded variation. (This functional was introduced by Hofbauer and Keller in the case of maps of the interval).

If x is a left and/or right limit of points of X, let

$$g(x, -) = \lim_{y \nearrow x} g(y)$$
$$g(x, +) = \lim_{y \searrow x} g(y).$$

Also let

$$f(x, -) = (fx, -) \text{ resp. } (fx, +)$$

if f is increasing (resp. decreasing) to the left of x and

$$f(x, +) = (fx, +) \text{ resp. } (fx, -)$$

if f is increasing (resp. decreasing) to the right of x.

We say that $(x, \pm)$ is a virtual periodic point if $f^m(x, \pm) = (x, \pm)$. A periodic point x of (minimal) period m may be accompanied either by a virtual periodic point of (minimal) period $2m$ or by at most two virtual periodic points of (minimal) period m.

---

[8]See the Appendix (Section 7).

If $x \in \text{Fix } f^m$, let

$$\Theta(x) = \left| \prod_{k=0}^{m-1} g\left(f^k x\right) \right|^{1/m}.$$

If $(x, \pm)$ is a virtual periodic point of period n, let

$$\Theta(x, \pm) = \left| \prod_{k=0}^{n-1} g\left(f^k(x, \pm)\right) \right|^{1/n}.$$

If ρ is an f-invariant probability measure on X, let

$$\Theta(\rho) = \exp \int \rho(dx) \log |g(x)|.$$

PROPOSITION 3.1. *Define*

$$\Theta = \lim_{m \to \infty} \sup_{x \in X} \left| \prod_{k=0}^{m-1} g\left(f^k x\right) \right|^{1/m}$$

$$= \exp \inf_m \sup_{x \in X} \frac{1}{m} \sum_{k=0}^{m-1} \log \left| g\left(f^k x\right) \right|.$$

Then $\Theta = \max\{\Theta_{\text{per}}, \Theta_{\text{vir}}, \Theta_{\text{erg}}\}$ where Θ_{per} is the sup *of $\Theta(x)$ over periodic points, Θ_{vir} is the* sup *of $\Theta(x, \pm)$ over virtual periodic points, and Θ_{erg} is the* sup *of $\Theta(\rho)$ over nonatomic f-ergodic measures ρ.*

First note that if

$$C(m) = \sup_{x \in X} \sum_{k=0}^{m-1} \log \left| g\left(f^k x\right) \right|$$

we have

$$C(m+n) \le C(m) + C(n)$$

hence

$$\lim_{m \to \infty} \frac{1}{m} C(m) = \inf_m \frac{1}{m} C(m)$$

which justifies the definition of Θ.

Taking x periodic, or tending to a periodic point, or ρ-almost any x, we find that $\Theta \ge \Theta_{\text{per}}, \Theta_{\text{vir}}, \Theta_{\text{erg}}$. It remains to show that $\Theta \le \max\{\Theta_{\text{per}}, \Theta_{\text{vir}}, \Theta_{\text{erg}}\}$, and it suffices to consider the case $\Theta \ne 0$. Let then $x(\alpha), m(\alpha)$ be such that $m(\alpha) \to \infty$ and

$$\frac{1}{m(\alpha)} \sum_{k=0}^{m(\alpha)-1} \log \left| g\left(f^k x(\alpha)\right) \right| \to \log \Theta.$$

By going to a subsequence we may assume that $\rho(\alpha) = 1/m(\alpha) \times \sum_{k=0}^{m(\alpha)-1} \delta_{f^k x(\alpha)}$ tends vaguely to a probability measure ρ on X. We write $\rho = \rho_0 + \rho_1$ where ρ_0 is atomic (hence carried by periodic orbits) and ρ_1 is nonatomic.

We shall now choose $\rho_0(\alpha)$, $\rho_1(\alpha)$ such that $\rho(\alpha) = \rho_0(\alpha) + \rho_1(\alpha)$ and $\rho_0(\alpha) \to \rho_0$, $\rho_1(\alpha) \to \rho_1$ vaguely. If $\Theta_{\mathrm{per}} = \Theta_{\mathrm{vir}} = 0$ then $\rho_0 = 0$ and we also take $\rho_0(\alpha) = 0$. Otherwise let $\delta > 0$; we take $\rho_0(\alpha) = \frac{1}{m(\alpha)} \sum^* \delta_{f^k x(\alpha)}$ where, for given α, $\sum^*$ is over long stretches of values of k such that $f^k x(\alpha)$ is close to a finite set of periodic orbits; in particular we may suppose that

$$\frac{1}{m(\alpha)} \sum^* \log \left| g\left(f^k x(\alpha) \right) \right| \le \|\rho_0(\alpha)\| \left(\log \max\{\Theta_{\mathrm{per}}, \Theta_{\mathrm{vir}}\} + \delta \right).$$

We only have to consider the case $\rho_1 \ne 0$. Suppose first that $\rho_1(\log|g|) = -\infty$. Given $N > 0$ we may choose $\varepsilon > 0$ such that if $|g|_\varepsilon(x) = \max\{|g(x)|, \varepsilon\}$ then $\rho_1(\log|g|_\varepsilon) < -N$. Note that ρ_1 gives zero measure to the set of discontinuities of $\log|g|_\varepsilon$ which is a bounded function of bounded variation, hence $\rho_1(\alpha)(\log|g|_\varepsilon) \to \rho_1(\log|g|_\varepsilon) \le -N$. Therefore

$$\log \Theta = \lim \rho(\alpha)(\log|g|)$$
$$\le \|\rho_0\| \left(\log \max\{\Theta_{\mathrm{per}}, \Theta_{\mathrm{vir}}\} + \delta \right) - N$$

for all $N \ge 0$, hence $\Theta = 0$ contrary to our assumption.

We have thus $\rho_1(\log|g|) > -\infty$ and we may choose $\varepsilon > 0$ such that

$$\rho_1(\log|g|_\varepsilon) \le \rho_1(\log|g|) + \|\rho_1\| \cdot \delta.$$

Since ρ_1 gives zero measure to the set of discontinuities of $\log|g|_\varepsilon$ we have $\rho_1(\alpha)\ (\log|g|_\varepsilon) \to \rho_1(\log|g|_\varepsilon)$ hence

$$\log \Theta = \lim \rho(\alpha)(\log|g|)$$
$$\le \|\rho_0\| \left(\log \max\{\Theta_{\mathrm{per}}, \Theta_{\mathrm{vir}}\} + \delta \right) + \|\rho_1\| \left(\log \Theta(\rho_1') + \delta \right)$$

where $\rho_1' = \rho_1/\|\rho_1\|$. The pointwise ergodic theorem, and Bogoliubov-Krylov theory, imply that $\Theta(\rho_1') \le \Theta(\rho')$ for some ergodic ρ', and since $\Theta(\rho') \le \max\{\Theta_{\mathrm{per}}, \Theta_{\mathrm{erg}}\}$, we have finally $\log \Theta \le \delta + \log \max\{\Theta_{\mathrm{per}}, \Theta_{\mathrm{vir}}, \Theta_{\mathrm{erg}}\}$ proving the proposition.

PROPOSITION 3.2. *Let Θ be associated with the system (X, f, g) and $\widehat{\Theta}$ similarly associated with the system $(\widehat{X}, \hat{f}, \hat{g})$ occurring in Propositions 2.1, 2.3, 2.4, 2.7. For Propositions 2.1, 2.3 we have $\widehat{\Theta} = \Theta$. For Proposition 2.4 we may choose $\hat{g}$ so that $\widehat{\Theta} \le \Theta$, and for Proposition 2.7 we have $\widehat{\Theta} \le \Theta$.*

In the situation of Proposition 2.1, passing from the first system to the second may replace some periodic orbits by others with the same Θ, leaving virtual periodic points and ergodic measures unchanged, therefore $\widehat{\Theta} = \Theta$.

In the situation of Proposition 2.3, the choice of $\hat{g}$ is such that if $\xi \in \text{Fix } \hat{f}^m$ we have $\widehat{\Theta}(\xi) = 0$ unless $\xi = \pi x$ with $x \in \text{Fix } f^m$, in which case $\widehat{\Theta}(\xi) = \Theta(\xi)$. If $(\xi, \pm)$ is a virtual periodic point for $\hat{f}$, we have $\widehat{\Theta}(\xi, \pm) = 0$ unless $\xi = \pi x$ and x is f-periodic, in which case either $(x, \pm)$ is a virtual periodic point for f and $\widehat{\Theta}(\xi, \pm) = \Theta(x, \pm)$, or $\widehat{\Theta}(\xi, \pm) = 0$. Finally, if $\hat{\rho}$ is $\hat{f}$-ergodic, and $\widehat{\Theta}(\hat{\rho}) > 0$, then $\hat{\rho}$ is carried by points of the form πx, so that $\hat{\rho} = \pi \rho$ where ρ is f-ergodic, and $\widehat{\Theta}(\hat{\rho}) = \Theta(\rho)$. In conlusion $\widehat{\Theta} = \Theta$.

In the situation of Proposition 2.4, we first have to extend the definition of $\hat{g}$ to all of $\widehat{X}$. If ξ is $\hat{f}$-periodic of period m, then the closed interval $\pi^{-1}\xi$ is mapped into itself by f^m. Choose a point $x \in \pi^{-1}\xi \cap \text{Fix } f^m$ if this set is not empty and define $\hat{g}(\hat{f}^k \xi) = g(f^k x)$ for $k \geq 0$. If $\pi^{-1}\xi \cap \text{Fix } f^m = \phi$ then $\xi \in \text{Fix}^- \hat{f}^m$ and there exists $x \in \pi^{-1}\xi \cap \text{Fix } f^{2m}$; we define then

$$\hat{g}(\hat{f}^k \xi) = \left(g(f^k x) \cdot g(f^{m+k} x) \right)^{1/2}$$

for $k \geq 0$, where the square root is chosen such that $\hat{g}(\hat{f}^k \xi)$, $g(f^k x)$, $g(f^{m+k} x)$ lie in the same half complex plane with boundary through the origin. This definition implies that

(3.1)
$$\left| \hat{g}(\hat{f}^k x) - g(f^k x) \right| + \left| g(f^{m+k} x) - \hat{g}(\hat{f}^k x) \right| \leq 2 \left| g(f^{m+k} x) - g(f^k x) \right|$$

[by an easy geometric argument using similarity of triangles]. Having defined $\hat{g}(\xi)$ when ξ is periodic we take $\hat{g}(\xi) \in g(\pi^{-1}\xi)$ for other ξ. (The above definition is compatible with those of Proposition 2.4, and Remark 2.5 (1). In view of (3.1), $\text{Var } \hat{g} \leq 2 \text{ Var } g$.

By construction $\widehat{\Theta}_{\text{per}} \leq \Theta_{\text{per}}$. If $\hat{\rho}$ is $\hat{f}$-ergodic nonatomic, it gives probability 0 to the countable set $\{\xi : \text{card } \pi^{-1}\xi > 1\}$, therefore there is an f-ergodic measure ρ with $\pi\rho = \hat{\rho}$ and $\widehat{\Theta}(\hat{\rho}) = \Theta(\rho)$, hence $\widehat{\Theta}_{\text{erg}} \leq \Theta_{\text{erg}}$. Let $(\xi, -)$ be a virtual periodic point of period m for $\hat{f}$, and $\xi_\alpha \nearrow \xi$; if $x_\alpha \in \pi^{-1}\xi_\alpha$ and $x_\alpha \nearrow x$, then $x \in \text{Fix } f^m$ and $\lim \hat{g}(\xi_\alpha) = \lim g(x_\alpha)$ so that $\widehat{\Theta}(\xi, -) = \Theta(x, -)$, and similarly with $(\xi, +)$. Therefore $\widehat{\Theta}_{\text{vir}} \leq \Theta_{\text{vir}}$. So finally $\widehat{\Theta} \leq \Theta$.

In the situation of Proposition 2.7, $\widehat{\Theta}_{\text{vir}} = \Theta_{\text{vir}}$ and $\widehat{\Theta}_{\text{erg}} = \Theta_{\text{erg}}$. The values of $\widehat{\Theta}$ associated with periodic points of the new system are values of Θ associated with the periodic or virtual periodic points of the old system. Therefore $\widehat{\Theta} \leq \Theta$.

4. The transfer operator $\mathcal{L}$.

As usual we consider a system (X, f, g) consisting of a compact subset X of $\mathbb{R}$, a piecewise monotone map f, a function g of bounded variation, and we choose a minimal cover $(J_1, \ldots, J_N)$ associated with f. We call *transfer operator* the operator $\mathcal{L}$ on the Banach space $\mathcal{B}$ (of functions $X \to \mathbb{C}$ of bounded variation) defined by

$$(\mathcal{L}\Phi)(x) = \sum_{y:fy=x} g(y)\Phi(y)$$

and we write[9]

$$R = \lim_{m \to \infty} \left(\|\mathcal{L}^m\|_0 \right)^{1/m}.$$

i.e., R is the spectral radius of $\mathcal{L}$ acting on bounded functions $X \to \mathbb{C}$.

THEOREM 4.1.   (a) *The spectral radius of $\mathcal{L}$, acting on $\mathcal{B}$, is $\geq \Theta$ and $\leq R$.*
  (b) *The essential spectral radius of $\mathcal{L}$ is $\leq \Theta$.*
  (c) *If $g \geq 0$, the spectral radius of $\mathcal{L}$ is R. If furthermore $\Theta < R$, then R is an eigenvalue of $\mathcal{L}$, and there is a corresponding eigenfunction $\Phi_0 \geq 0$.*

We shall see in Section 6 that $R \leq \max\left(\Theta, \exp P(\log|g|) \right)$ where P denotes the pressure.

PROOF OF PART (a). For each $m > 0$ choose y_m such that

$$\left| \prod_{k=0}^{m-1} g(f^k y_m) \right| \geq \frac{1}{2} \sup_y \left| \prod_{k=0}^{m-1} g(f^k y) \right|$$

and let Φ_m take the value 1 at y_m, 0 elsewhere. The spectral radius of $\mathcal{L}$ is then

$$\lim_{m \to \infty} \|\mathcal{L}^m\|^{1/m} \geq \lim \left(\frac{1}{2} \|\mathcal{L}^m \Phi_m\| \right)^{1/m}$$

$$= \lim \left| \prod_{k=0}^{m-1} g(f^k y_m) \right|^{1/m} \geq \lim \left(\frac{1}{2} \sup_y \left| \prod_{k=0}^{m-1} g(f^k y) \right| \right)^{1/m} = \Theta$$

---

[9]Elsewhere (Ruelle [42]) we have used R to denote the spectral radius of $|\mathcal{L}|$ acting on bounded functions, where $|\mathcal{L}|$ is the operator obtained when g is replaced by $|g|$ in the definition of $\mathcal{L}$.

and we also have

$$R = \lim_{m\to\infty}\left(\|\mathcal{L}^m\|_0\right)^{1/m} \geq \lim\left(\|\mathcal{L}^m\Phi_m\|_0\right)^{1/m}$$

$$= \lim\left|\prod_{k=0}^{m-1} g\left(f^k y_m\right)\right|^{1/m} = \Theta.$$

Let $\tilde{J}_i \subset J_i$ be such that $(\tilde{J}_1,\dots,\tilde{J}_N)$ is a partition of X, and let ψ_i be the inverse of $f|\tilde{J}_i$. Writing

$$\varphi_{i_1\dots i_k}(x) = g(\psi_{i_1}x)\cdot g(\psi_{i_2}\psi_{i_1}x)\cdot\cdots\cdot g(\psi_{i_k}\dots\psi_{i_1}x)$$

we have then

$$\mathcal{L}^m\Phi(x) = \sum_{i_1\dots i_m}\varphi_{i_1\dots i_m}(x)\Phi(\psi_{i_m}\dots\psi_{i_1}x)$$

$$\operatorname{Var}\mathcal{L}^m\Phi = \sum_{k=1}^m \sup_x\left|g\left(f^{k-2}x\right)\dots g(fx)g(x)\right|\cdot\sum_{i_k}\operatorname{Var}(g\circ\psi_{i_k})$$

$$\sup_y\left|\sum_{i_{k+1}\dots i_m}\varphi_{i_{k+1}\dots i_m}(\psi_{i_k}y)\Phi(\psi_{i_m}\dots\psi_{i_k}y)\right|$$

$$+ \sup_x\left|g\left(f^{m-1}x\right)\dots g(fx)g(x)\right|\cdot\operatorname{Var}\Phi.$$

Therefore if $\bar{\Theta} > \Theta$, $\bar{R} > R$, there are $C, C' > 0$ such that

$$\operatorname{Var}\mathcal{L}^m\Phi \leq C\left[\sum_i\operatorname{Var} g\circ\psi_i\cdot\sum^{m_{k=1}}\bar{\Theta}^{k-1}\bar{R}^{m-k}\|\Phi\|_0 + \bar{\Theta}^m\operatorname{Var}\Phi\right]$$

$$\leq (m+1)C'\left(\max(\bar{\Theta},\bar{R})\right)^m\cdot\operatorname{Var}\Phi$$

hence

$$\lim_{m\to\infty}\|\mathcal{L}^m\|^{1/m} \leq \max(\Theta,R) = R.$$

PROOF OF PART (b). Using the notation of the proof of part (a), let $x_{i_1\dots i_m}$ be in the domain of definition of $\psi_{i_m}\circ\cdots\circ\psi_{i_1}$ (if this is not empty) and define

$$(K_m\Phi)(x) = \sum_{i_1\dots i_m}\varphi_{i_1\dots i_m}(x)\Phi(\psi_{i_m}\dots\psi_{i_1}x_{i_1\dots i_m}).$$

The operator K_m has finite rank, and if we prove

$$(4.1)\qquad\qquad \limsup_{m\to\infty}\|\mathcal{L}^m - K_m\|^{1/m} \leq \Theta$$

then Nussbaum's essential spectral radius formula [**31**] implies that the essential spectral radius of $\mathcal{L}$ is $\leq \Theta$. We follow the calculation of part (a), with $\mathcal{L}^m\Phi$ replaced by $\mathcal{L}^m\Phi - K_m\Phi$. Note that

$$\sum_{i_{k+1}\cdots i_m} \left\| \Phi \circ \psi_{i_m} \circ \cdots \circ \psi_{i_k} - \Phi\left(\psi_{i_m} \circ \cdots \circ \psi_{i_k}(\psi_{i_{k-1}} \cdots \psi_{i_1} x_{i_1 \ldots i_m})\right) \right\|_0$$

$$\leq \text{var } \Phi$$

so that

$$\text{Var}(\mathcal{L}^m\Phi - K_m\Phi)$$

$$\leq C\left[\sum_i \text{Var } g \circ \psi_i \cdot \sum_{k=1}^m \bar{\Theta}^{k-1} \cdot \bar{\Theta}^{m-k} \cdot \text{var } \Phi + \bar{\Theta}^m \cdot 2\,\text{Var } \Phi\right]$$

$$\leq (m+1)C'\bar{\Theta}^m \text{ Var } \Phi$$

which implies (4.1).

PROOF OF PART (c). Since $g \geq 0$ we have

$$\begin{aligned}
\text{(4.2)} \quad \lim_{m\to\infty} \|\mathcal{L}^m\|^{1/m} &\geq \lim_{m\to\infty} \left(\text{Var}(\mathcal{L}^m 1)\right)^{1/m} \\
&\geq \lim_{m\to\infty} \left(\|\mathcal{L}^m 1\|_0\right)^{1/m} = \lim_{m\to\infty} \left(\|\mathcal{L}^m\|_0\right)^{1/m} = R
\end{aligned}$$

so that the spectral radius of $\mathcal{L}$ is $\geq R$. Therefore, part (a) of the theorem shows that the spectral radius of $\mathcal{L}$ is equal to R.

We assume now that $\Theta < R$ and prove that R is an eigenvalue of $\mathcal{L}$, and has an eigenfunction $\Phi_R \geq 0$. We may write

$$1 = \Psi + \sum_j \Psi_j$$

where, for each j, Ψ_j is in the generalized eigenspace of an eigenvalue λ_j of $\mathcal{L}$, with $|\lambda_j| = R$; furthermore

$$\lim_{m\to\infty} \frac{\text{Var } \mathcal{L}^m\Psi}{r^m} = 0$$

with $0 < r < R$. In view of (4.2)

$$\lim_{m\to\infty} \frac{1}{m} \log \text{Var}(\mathcal{L}^m 1) = \log R$$

and therefore the Ψ_j do not all vanish.

Writing the restriction of $\mathcal{L}$ to the generalized eigenspaces corresponding to the λ_j in Jordan normal form we see that there is an integer $k \geq 0$ such that

$$\lim_{m \to \infty} \frac{1}{\lambda_j^m m^k} \, \mathcal{L}^m \Psi_j = \Phi_j$$

$$\mathcal{L}\Phi_j = \lambda_j \Phi_j$$

for all j, and $\Phi_j \neq 0$ for some j. We have

$$0 \leq \frac{\mathcal{L}^m 1}{R^m m^k} = \frac{\mathcal{L}^m \Psi}{R^m m^k} + \sum_j \left(\frac{\lambda_j}{R}\right)^m \frac{\mathcal{L}^m \Psi_j}{\lambda_j^m m^k}$$

and therefore

(4.3) $$\sum_j \left(\frac{\lambda_j}{R}\right)^m \Phi_j \geq -\varepsilon(m)$$

where $0 \leq \varepsilon(m) \to 0$ when $m \to \infty$. Note that the sum in the left-hand side is finite and that $|\lambda_j/R| = 1$ for all j.

Let $\langle \ldots \rangle_m$ denote the average over $m \in \mathbb{Z}$; (4.3) yields then

$$\left\langle (1 \pm \cos \alpha m) \sum_j \left(\frac{\lambda_j}{R}\right)^m \Phi_j \right\rangle_m \geq 0$$

$$\left\langle (1 \pm \sin \alpha m) \sum_j \left(\frac{\lambda_j}{R}\right)^m \Phi_j \right\rangle_m \geq 0.$$

If $\lambda_j \neq R$ for all j we would have

$$\left\langle \sum_j \left(\frac{\lambda_j}{R}\right)^m \Phi_j \right\rangle_m = 0$$

hence

$$\left\langle e^{i\alpha m} \sum_j \left(\frac{\lambda_j}{R}\right)^m \Phi_j \right\rangle_m = 0$$

for all real α. But then $\Phi_j = 0$ for all j: a contradiction. Therefore R is an eigenvalue, say $\lambda_0 = R$. Furthermore Φ_0 does not vanish identically, and $\Phi_0 = \langle \sum_j \left(\frac{\lambda_j}{R}\right)^m \Phi_j \rangle_m \geq 0$.

LEMMA 4.2. *Let (U_α) be a family of disjoint intervals of X (not necessarily closed), and write $Y = \bigcup_\alpha U_\alpha$. We assume that each U_α is contained in some J_i and that either*

   (a_1) *fU_α is contained in another interval U_β of the family, or*
   (a_2) *g vanishes on U_α.*
We also assume that for each U_α either
   (b_1) *if $U_\beta < U_\alpha$ there exists $x_{\alpha\beta} \in X \backslash Y$ with $U_\beta < x_{\alpha\beta} < U_\alpha$, or*

(b_2) *if $U_\beta > U_\alpha$ there exists $y_{\alpha\beta} \in X \backslash Y$ with $U_\alpha < x_{\alpha\beta} < U_\beta$.*

The conditions (a) *imply that $\mathcal{L}\mathcal{B}_Y \subset \mathcal{B}_Y$. An operator $\mathcal{L}_{\backslash Y}$ on $\mathcal{B}_{\backslash Y}$ is thus defined such that $\mathcal{L}_{\backslash Y}\omega = \omega\mathcal{L}$ if $\omega : \mathcal{B} \to \mathcal{B}_{\backslash Y}$ is the quotient map.*

The conditions (b) *imply that $\mathcal{L}|\mathcal{B}_Y$ has spectral radius $\leq \Theta$. Therefore if $|\lambda| > \Theta$ and E^λ, $E^\lambda_{\backslash Y}$ are the generalized eigenspaces of $\mathcal{L}$, $\mathcal{L}_{\backslash Y}$ to the eigenvalue λ, then $\omega : \mathcal{B} \to \mathcal{B}_{\backslash Y}$ induces a bijection $E^\lambda \to E^\lambda_{\backslash Y}$.*

Suppose that Ψ vanishes outside of Y, then $(\mathcal{L}\Psi)(x)$ vanishes unless $x = fy$ with $y \in U_\alpha$ and $g(y) \neq 0$, hence $x \in U_\beta \subset Y$ by the conditions (a). Therefore $\mathcal{L}\mathcal{B}_Y \subset \mathcal{B}_Y$ and $\mathcal{L}_{\backslash Y}$ is well defined.

The conditions (b) imply that if $\Psi \in \mathcal{B}_Y$

$$\text{Var}\, \Psi \leq \sum_\alpha \text{Var}(\Psi | U_\alpha) \leq 2\, \text{Var}\, \Psi.$$

[The second inequality is because, in evaluating

$$\text{Var}\, \Psi = \lim \left(|\Psi(a_0)| + \sum |\Psi(a_i) - \Psi(a_{i-1})| + |\Psi(a_n)| \right)$$

we may assume that each stretch of $a_i's$ in a given U_α is immediately preceded or followed by some a_j with $\Psi(a_j) = 0$].

We have thus, if $\Psi \in \mathcal{B}_Y$,

$$\text{Var}\, \mathcal{L}^m\Psi \leq \sum_\beta \text{Var}(\mathcal{L}^m\Psi | U_\beta)$$

$$\leq \sum_\alpha \text{Var}\left(\mathcal{L}^m(\chi_{U_\alpha}\Psi) \right)$$

and, if ψ_i is the inverse of $f|J_i$,

$$\mathcal{L}^m(\chi_{U_\alpha}\Psi) = (g \circ \psi_{i_1}) \ldots (g \circ \psi_{i_1} \circ \cdots \circ \psi_{i_m})\left((\chi_{U_\alpha}\Psi) \circ \psi_{i_1} \circ \cdots \circ \psi_{i_m} \right).$$

Given $\varepsilon > 0$ we may bound a product of k factors g on an orbit of f by $C(\Theta + \varepsilon)^k$, therefore

$$\text{Var}\left(\mathcal{L}^m(\chi_{U_\alpha}\Psi) \right) \leq mC^2(\Theta + \varepsilon)^{m-1} \cdot \text{Var}\, g \cdot \text{Var}(\Psi | U_\alpha)$$

and finally

$$\text{Var}\, \mathcal{L}^m\Psi \leq mC^2(\Theta + \varepsilon)^{m-1} \cdot \text{Var}\, g \cdot 2\, \text{Var}\, \Psi.$$

Therefore

$$\| \mathcal{L}^m | \mathcal{B}_Y \| \leq 2mC^2(\Theta + \varepsilon)^{m-1} \cdot \text{Var}\, g$$

and the spectral radius of $\mathcal{L}|\mathcal{B}_Y$ is $\leq \Theta$, from which the lemma results.

PROPOSITION 4.3. *Suppose that $\mathcal{L}$, Y, Θ satisfy the conditions of Lemma 4.2, and similarly $\widehat{\mathcal{L}}$, $\widehat{Y}$, $\widehat{\Theta}$, with $\widehat{\Theta} \leq \Theta$. If there is an isomorphism $\mathcal{B}_{\backslash Y} \simeq \widehat{\mathcal{B}}_{\backslash \widehat{Y}}$ identifying $\widehat{\mathcal{L}}_{\backslash Y}$ with $\widehat{\mathcal{L}}_{\backslash \widehat{Y}}$ we say that the transfer operators $\mathcal{L}$ and $\widehat{\mathcal{L}}$ are Θ-equivalent: their eigenvalues λ with $|\lambda| > \Theta$ are the same and the generalized eigenspaces E^λ, $\widehat{E}^\lambda$ are in natural correspondence. We extend the notion of Θ-equivalence by transitivity.*

The constructions of Proposition 2.1 (producing a partition), Proposition 2.3 (producing a Markov partition), Proposition 2.4 when $\widehat{\Theta} \leq \Theta$ (producing a generating partition), and Proposition 2.7 (producing $\hat{g}$ continuous at periodic points) yield Θ-equivalences $\mathcal{L} \sim \widehat{\mathcal{L}}$.

In the case of Proposition 2.4 we shall at first consider the subcase where $(J_1, \ldots, J_N)$ is a Markov partition. The conditions of applicability of Lemma 4.2 have already been checked in Propositions 2.1, 2.3 (with $Y = \phi$), 2.4 (in the Markov subcase), and 2.7. Furthermore it is clear by construction that the isomorphisms $\mathcal{B}_{\backslash Y} \simeq \widehat{\mathcal{B}}_{\backslash \widehat{Y}}$ defined by $\Phi \mapsto \Phi \circ \hat{\pi}$ (Proposition 2.1), $\widehat{\Phi}|X \longleftarrow \widehat{\Phi}$ (Proposition 2.3), $\widehat{\Phi} \circ \pi \longleftarrow \widehat{\Phi}$ (Proposition 2.4) and $\Phi \mapsto \Phi \circ \hat{\pi}$ (Proposition 2.7) identify $\mathcal{L}_{\backslash Y}$ with $\widehat{\mathcal{L}}_{\backslash \widehat{Y}}$.

Apart from the Markov restriction in the case of Proposition 2.4 we have thus established the announced Θ-equivalence $\mathcal{L} \sim \widehat{\mathcal{L}}$.

For the general case of Proposition 2.4 we use the fact that one can apply Propositions 2.3 and 2.4 in either order, getting almost the same result, as noted in Remark 2.5 (3). If we mark by $(*)$ or $(\hat{\ })$ the application of Propositions 2.3 and 2.4 respectively, we have transfer operators $\mathcal{L}$, $\mathcal{L}^*$, $\widehat{\mathcal{L}}$, $\widehat{\mathcal{L}}^*$ and $(\mathcal{L}^*)\hat{\ }$ with Θ-equivalences $\mathcal{L} \sim \mathcal{L}^* \sim (\mathcal{L}^*)\hat{\ }$, $\widehat{\mathcal{L}} \sim \widehat{\mathcal{L}}^*$. But by Proposition 2.3 and Remark (3) after Proposition 2.4 we also have $(\mathcal{L}^*)\hat{\ } \sim \widehat{\mathcal{L}}^*$. By transitivity we have thus finally $\mathcal{L} \sim \widehat{\mathcal{L}}$.

Remarks 4.4. Suppose that X is a Cantor set. Let $\mathcal{B}_\infty = \{\Phi \in \mathcal{B} : \{x : \Phi(x) \neq 0\}$ is countable$\}$ and define $\mathcal{L}_{\backslash \infty}\omega = \omega\mathcal{L}$ where

$$\omega : \mathcal{B} \to \mathcal{B}_{\backslash \infty} = \mathcal{B}/\mathcal{B}_\infty$$

is the quotient map. Then the eigenvalues λ with $|\lambda| > \Theta$ coincide for $\mathcal{L}$ and $\mathcal{L}_{\backslash \infty}$, and ω induces a bijection $E^\lambda \to E^\lambda_{\backslash \infty}$ of the corresponding generalized eigenspaces.

Fixing $\Psi \in \mathcal{B}_\infty$, let $Z = \{x : \Psi(x) \neq 0\}$ and $Y = \bigcup_{m \geq 0} f^m Z$. We may apply Lemma 4.2 because every point of the countable set Y is a limit of points of $X \backslash Y$ (X is a Cantor set). Therefore (as in the proof of Lemma 4.2)

$$\|\mathcal{L}^m \Psi\| \leq 2mC^2(\Theta + \varepsilon)^{m-1} \operatorname{Var} g \cdot \|\Psi\|$$

hence the spectral radius of $\mathcal{L}|\mathcal{B}_\infty$ is $\leq \Theta$, and this justifies the Remark.

Remarks 4.5. Suppose that X is a Cantor set, that $\mathrm{Var}\,\tilde{g} < \infty$, that $\{x : \tilde{g}(x) \neq g(x)\}$ is countable, and that

$$\lim_{m \to \infty} \sup_x \left| \prod_{k=0}^{m-1} \tilde{g}(f^k x) \right|^{1/m} \leq \Theta.$$

Then the transfer operator $\tilde{\mathcal{L}}$ associated with $\tilde{g}$ is Θ-equivalent to $\mathcal{L}$. We use Remark 4.4 and notice that $\mathcal{L}_{\backslash\infty} = \tilde{\mathcal{L}}_{\backslash\infty}$.

5. Zeta functions.

At the level of generality considered here, the study of zeta functions for piecewise monotone maps was initiated by Baladi and Keller [4] who discussed the case where X is an interval of $\mathbb{R}$ and the minimal cover $(J_1, \ldots, J_N)$ is generating. Their proof simplifies if one assumes that $(J_1, \ldots, J_N)$ is a generating Markov partition (and X a Cantor set); we shall examine that case first, and then use the machinery developed earlier to handle more general cases (and recover in particular the Baladi-Keller theorem).

PROPOSITION 5.1. *Let X be a Cantor subset of $\mathbb{R}$, $f : X \to X$ a piecewise monotone map, g a function of bounded variation, and $(J_1, \ldots, J_N)$ a partition associated with f.*

We assume that $(J_1, \ldots, J_N)$ is generating, that $fJ_i = X$ for $i = 1,$ $\ldots$, N (in particular $(J_1, \ldots, J_N)$ is a Markov partition), and we define a zeta function by

$$\zeta(z) = \exp \sum_{m=1}^{\infty} \frac{z^m}{m} \sum_{x \in \mathrm{Fix}\,f^m} \prod_{k=0}^{m-1} g(f^k x).$$

Then $1/\zeta(z)$ is analytic for $|z| < \Theta^{-1}$ and its zeros there are the inverses λ^{-1} of the eigenvalues λ of the transfer matrix $\mathcal{L}$, with the same multiplicity.

The original idea of the proof is due to Haydn [19], who worked with Hölder continuous g. Haydn's argument was adapted to the present situation by Baladi and Keller.

If Λ_+ is the set of sequences $\xi = (\xi_0, \xi_1, \ldots)$ with $\xi_i \in \{1, \ldots, N\}$, and τ is the shift : $\tau\xi = (\xi_1, \xi_2, \ldots)$, we may identify (X, f) with (Λ_+, τ) by taking $x \mapsto (\xi_0, \xi_1, \ldots)$ such that $f^k x \in J_{\xi_k}$. We shall freely change notation in accordance with this identification.

We write

$$\zeta_m = \sum_{\xi \in \mathrm{Fix}\,\tau^m} \prod_{k=0}^{m-1} g(\tau^k \xi)$$

so that

$$1/\zeta(z) = \exp - \sum_{m=1}^{\infty} \frac{1}{m} \zeta_m z^m.$$

According to Theorem 4.1 (b) we may choose $\bar{\Theta}$, with $0 < \bar{\Theta} - \Theta$ arbitrarily small, such that there is no eigenvalue λ of the transfer operator $\mathcal{L}$ with $|\lambda| = \bar{\Theta}$, and finitely many (say M) with $|\lambda| > \bar{\Theta}$. The projection $\mathcal{P}$ corresponding to the part of the spectrum of $\mathcal{L}$ in $\{\lambda : |\lambda| < \bar{\Theta}\}$ is a bounded operator in $\mathcal{B}$ and we may write

$$\mathcal{L}\Phi = \sum_{j=1}^{M} \lambda_j \sum_{\alpha\beta} S_{j\alpha} \cdot (L_j)_{\alpha\beta} \sigma_{j\beta}(\Phi) + \mathcal{P}\mathcal{L}\Phi$$

$$\mathcal{L}^m\Phi = \sum_{j=1}^{M} \lambda_j^m \sum_{\alpha\beta} S_{j\alpha} \cdot \left(L_j^m\right)_{\alpha\beta} \sigma_{j\beta}(\Phi) + \mathcal{P}\mathcal{L}^m\Phi$$

where $\Theta < |\lambda_j| \leq R$, $(S_{j\alpha})$ and $(\sigma_{j\alpha})$ are dual bases of the generalized eigenspaces of $\mathcal{L}$ and $\mathcal{L}^*$ corresponding to the eigenvalue λ_j, and the matrices L_j may be assumed to be in Jordan normal form (L_j^m is the m-th power of the matrix L_j). We let m_j be the multiplicity of λ_j, i.e., $m_j = \operatorname{tr} L_j$.

Fixing $m > 0$ we let $\sum_\eta$ be the sum over words η of length m, i.e., elements of $\{1, \ldots, N\}^m$. We denote by η^* the periodic concatenation[10] $\eta^* = \eta \vee \eta \vee \cdots \in \Lambda_+$. Let $\chi_\eta \in \mathcal{B}$ be such that $\chi_\eta(\xi) = 1$ if ξ begins with the word η, $\chi_\eta(\xi) = 0$ otherwise. Then

$$(\mathcal{L}^m \chi_\eta)(\xi) = \prod_{k=0}^{m-1} g\left(\tau^k(\eta \vee \xi)\right).$$

Note that

$$\|\mathcal{L}^m \chi_\eta\|_0 \leq C_1 \bar{\Theta}^m$$

for some constant C_1, hence

$$\operatorname{Var} \mathcal{L}^m \chi_\eta \leq m \operatorname{Var} g \cdot C_1^2 \bar{\Theta}^{m-1}.$$

Since $\bar{\Theta}$ may be slightly changed we can get rid of the factor m and write

$$\operatorname{Var} \mathcal{L}^m \chi_\eta \leq C_2 \bar{\Theta}^m.$$

We have

$$\zeta_m = \sum_\eta \prod_{k=0}^{m-1} g\left(\tau^k \eta^*\right) = \sum_\eta \prod_{k=0}^{m-1} g\left(\tau^k(\eta \vee \eta^*)\right)$$

---

[10]The concatenation $\eta \vee \eta'$ of a word η of length m and a word η' of length m' is a word of length $m + m'$.

$$= \sum_\eta (\mathcal{L}^m \chi_\eta)(\eta^*)$$

$$= \sum_\eta \left[\sum_{j=1}^M \lambda_j^m \sum_{\alpha\beta} S_{j\alpha}(\eta^*) \cdot \left(L_j^m \right)_{\alpha\beta} \sigma_{j\beta}(\chi_\eta) + (\mathcal{PL}^m \chi_\eta)(\eta^*) \right]$$

$$= \sum_{j=1}^M \lambda_j^m \sum_{\alpha\beta} \left(L_j^m \right)_{\alpha\beta} \sigma_{j\beta} \left(\sum_\eta S_{j\alpha}(\eta^*) \chi_\eta \right) + \sum_\eta (\mathcal{PL}^m \chi_\eta)(\eta^*)$$

$$= \zeta_m^{(0)} + \zeta_m^{(1)} + \zeta_m^{(2)}$$

where

$$\zeta_m^{(0)} = \sum_{j=1}^M m_j \lambda_j^m$$

$$\zeta_m^{(1)} = \sum_{j=1}^M \lambda_j^m \sum_{\alpha\beta} \left(L_j^m \right)_{\alpha\beta} \sigma_{j\beta} \left(\sum_\eta S_{j\alpha}(\eta^*) \chi_\eta - S_{j\alpha} \right)$$

$$\zeta_m^{(2)} = \sum_\eta (\mathcal{PL}^m \chi_\eta)(\eta^*).$$

We have

$$\exp - \sum_{m=1}^\infty \frac{1}{m} \zeta_m^{(0)} z^m = \prod_{j=1}^M (1 - \lambda_j z)^{m_j}.$$

Therefore the proposition follows from Lemma 5.2 and Lemma 5.3 below.

LEMMA 5.2. *There is a constant C such that*

$$|\zeta_m^{(1)}| \leq C \bar{\Theta}^m.$$

We have the formula

$$\mathcal{L}^{*m} \sigma_{j\alpha} = \lambda_j^m \sum_\beta \left(L_j^m \right)_{\alpha\beta} \sigma_{j\beta}.$$

Therefore

$$\zeta_m^{(1)} = -\sum_{j=1}^M \lambda_j^m \sum_{\alpha\beta} \left(L_j^m \right)_{\alpha\beta} \sigma_{j\beta} \left(S_{j\alpha} - \sum_\eta S_{j\alpha}(\eta^*) \chi_\eta \right)$$

$$= -\sum_{j=1}^M \sum_\alpha (\mathcal{L}^{*m} \sigma_{j\alpha}) \left(S_{j\alpha} - \sum_\eta S_{j\alpha}(\eta^*) \chi_\eta \right)$$

$$= -\sum_{j=1}^M \sum_\alpha \sigma_{j\alpha} \left(\mathcal{L}^m \left(S_{j\alpha} - \sum_\eta S_{j\alpha}(\eta^*) \chi_\eta \right) \right)$$

$$= -\sum_{j=1}^{M}\sum_{\alpha}\sigma_{j\alpha}(\mathcal{L}^m S_{j\alpha} - K_m S_{j\alpha})$$

where K_m is an operator of finite rank defined by

$$K_m = \mathcal{L}^m E_m$$

$$(E_m\Phi)(\xi) = \sum_{\eta}\Phi(\eta^*)\chi_\eta(\xi).$$

If $\Psi_\eta = \chi_\eta \cdot (1 - E_m)\Phi = \chi_\eta \cdot \left(\Phi - \Phi(\eta^*)\right)$ we have

$$\sum_{\eta}\operatorname{Var}\Psi_\eta \le 3\operatorname{Var}\Phi$$

so that

$$\operatorname{Var}(\mathcal{L}^m\Phi - K_m\Phi) = \operatorname{Var}\left(\mathcal{L}_m(1 - E_m)\Phi\right) = \operatorname{Var}\sum_{\eta}\mathcal{L}^m\Psi_\eta$$

$$\le \sum_{\eta}\operatorname{Var}\mathcal{L}^m\chi_\eta \cdot \operatorname{Var}\Psi_\eta \le 3C_2\bar{\Theta}^m\operatorname{Var}\Phi$$

hence $\|\mathcal{L}^m - K_m\| \le 3C_2\bar{\Theta}^m$. Finally

$$\left|\zeta_m^{(1)}\right| \le C \cdot \bar{\Theta}^m$$

as announced.

LEMMA 5.3. *There is a constant C' such that*

$$\left|\zeta_m^{(2)}\right| \le C'\bar{\Theta}^m.$$

Fix $\tilde{\eta} \in \Lambda_+$. If η_k is a word of length k, we define

$$Y_{\eta_k} = \begin{cases} \mathcal{L}^k\chi_{\eta_k} - g(\eta_k \vee \tilde{\eta}) \cdot \mathcal{L}^{k-1}\chi_{\tau\eta_k} & \text{if } k \ge 2 \\ \mathcal{L}\chi_{\eta_k} & \text{if } k = 1 \end{cases}$$

where $\tau\eta_k$ is the word of length $k - 1$ obtained by application of the shift, and χ_{η_k} is the characteristic function of elements of Λ_+ with η_k as initial word. We have, if $k \ge 2$,

$$\operatorname{Var}Y_{\eta_k} = \operatorname{Var}\left[\left(g(\eta_k \vee \cdot) - g(\eta_k \vee \tilde{\eta})\right) \cdot \left(\mathcal{L}^{k-1}\chi_{\tau\eta_k}\right)(\cdot)\right]$$

$$\le \frac{3}{2}\operatorname{Var}g \cdot \operatorname{Var}\mathcal{L}^{k-1}\chi_{\tau\eta_k}$$

$$\le \frac{3}{2}\operatorname{Var}g \cdot C_2\bar{\Theta}^{k-1} = C_3\bar{\Theta}^k.$$

This inequality will be used in a moment.

Writing $\eta_m = \eta$ we have

$$\mathcal{L}^m \chi_\eta = \sum_{k=0}^{m-1} g(\eta \vee \tilde{\eta}) \ldots g\left(\tau^{k-1}\eta \vee \tilde{\eta}\right) Y_{\tau^k \eta}$$

hence

$$\zeta_m^{(2)} = \sum_\eta (\mathcal{P}\mathcal{L}^m \chi_\eta)(\eta^*)$$

$$= \sum_\eta \sum_{k=0}^{m-1} g(\eta \vee \tilde{\eta}) \ldots g\left(\tau^{k-1}\eta \vee \tilde{\eta}\right) \cdot (\mathcal{P}Y_{\tau^k \eta})(\eta^*)$$

$$= a + b$$

where

$$a = \sum_{k=0}^{m-1} \sum_\eta g(\eta \vee \tilde{\eta}) \ldots g\left(\tau^{k-1}\eta \vee \tilde{\eta}\right) \left[(\mathcal{P}Y_{\tau^k \eta})(\eta^*) - (\mathcal{P}Y_{\tau^k \eta})(\eta \vee \tilde{\eta})\right]$$

$$b = \sum_{k=0}^{m-1} \sum_\eta g(\eta \vee \tilde{\eta}) \ldots g\left(\tau^{k-1}\eta \vee \tilde{\eta}\right) \cdot (\mathcal{P}Y_{\tau^k \eta})(\eta \vee \tilde{\eta}).$$

Furthermore

$$|a| \leq \sum_{k=0}^{m-1} \sum_\eta \left|g(\eta \vee \tilde{\eta}) \ldots g\left(\tau^{k-1}\eta \vee \tilde{\eta}\right)\right| \cdot \mathrm{var}(\mathcal{P}Y_{\tau^k \eta}|\eta \vee \Lambda_+)$$

$$\leq \sum_{k=0}^{m-1} C_1 \bar{\Theta}^k \sum_{\eta_{m-k}} \mathrm{Var}\, \mathcal{P}Y_{\eta_{m-k}}$$

$$\leq \sum_{k=0}^{m-1} C_1 \bar{\Theta}^k \|\mathcal{P}\| \cdot C_3 \bar{\Theta}^{m-k}$$

$$= m\|\mathcal{P}\| \cdot C_1 C_2 \bar{\Theta}^m.$$

Also

$$b = \sum_{k=0}^{m-1} \sum_{\eta_{m-k}} (\mathcal{L}^k \mathcal{P}Y_{\eta_{m-k}})(\eta_{m-k} \vee \tilde{\eta})$$

hence

$$|b| \leq \sum_{k=0}^{m-1} \sum_{\eta_{m-k}} \|\mathcal{P}\mathcal{L}^k\| \, \mathrm{Var}\, Y_{\eta_{m-k}}$$

$$\leq \sum_{k=0}^{m-1} C_4 \bar{\Theta}^k \cdot C_3 \bar{\Theta}^{m-k}$$

$$= m C_3 C_4 \bar{\Theta}^m.$$

Since $\bar{\Theta}$ may be slightly changed we can get rid of the factor m and obtain

$$\left| \zeta_m^{(2)} \right| \leq |a| + |b| \leq C' \bar{\Theta}^m$$

as desired.

Representative set of periodic points. Starting from the dynamical system (X, f) and the minimal cover $(J_1, \ldots, J_N)$ we obtain by applying successively Propositions 2.1 and 2.4 first a system $(\widetilde{X}, \tilde{f})$ then a system $(\widehat{X}, \hat{f})$. The elements of $\widehat{X}$ are of the form $(\xi_k)_{k \geq 0}$, i.e., sequences of *symbols* from the alphabet $\{1, \ldots, N\}$, and $\hat{f}$ is the *shift*. When the orbit $\left(f^k x \right)_{k \geq 0}$ avoids the set $\{b_1, \ldots, b_s\}$ of common endpoints of intervals J_i, J_{i+1}, the point x has a well-defined image $\alpha x = (\xi_k)_{k \geq 0} \in \widehat{X}$ such that $f^k x \in J_{\xi_k}$. In particular if Per f, Per $\hat{f}$ are the sets of periodic points for f, $\hat{f}$, there are finite sets P_*, $\widehat{P}_*$ such that

$$\alpha(\text{Per } f \backslash P_*) = \text{Per } \hat{f} \backslash \widehat{P}_*.$$

We say that an f-invariant set $S \subset \text{Per } f$ is a *representative set of periodic points* if α induces a bijection

$$\beta : S \backslash \text{finite set} \rightarrow \text{Per } \hat{f} \backslash \text{finite set}$$

such that the $\hat{f}$-period of βx is equal to the f-period of x.

As in Section 1 we let $\text{Per}^{\pm}(f, m)$ be the set of points in $\text{Per } f \backslash P_*$ which are positive $(+)$ or negative $(-)$ periodic points with minimal period m. Similarly $\text{Per}^{\pm}(\hat{f}, m)$ is the set of points in $\text{Per } \hat{f}$ which are positive or negative with minimal period m. Note that α preserves the period m of x if and only if it preserves its sign $\pm$, but that we may have $x \in \text{Per}^{+}(f, 2n)$ and $\alpha x \in \text{Per}^{-}(\hat{f}, n)$. For a positive periodic point $\xi \in \text{Per}^{+}(\hat{f}, m) \backslash \widehat{P}_*$ there is always $x \in \alpha^{-1}\xi \cap \text{Per}^{+}(f, m)$. If X is an interval of $\mathbb{R}$, Corollary 2.6 shows that for a negative periodic point $\xi \in \text{Per}^{-}(\hat{f}, m) \backslash \widehat{P}_*$ there is $x \in \alpha^{-1}\xi \cap \text{Per}^{-}(f, m)$, and x is unique. Therefore if X is an interval of $\mathbb{R}$ there is always a representative set of periodic points. This is part (a) of the following proposition.

PROPOSITION 5.4. *Let X be a compact subset of $\mathbb{R}$, f a piecewise monotone map of X, and $(J_1, \ldots, J_N)$ an associated minimal cover*[11].

    (a) *If X is an interval of $\mathbb{R}$ there is always a representative set S of periodic points.*

    (b) *In the following cases Per f itself is a representative set of periodic points:*

---

[11] As noted in Remark 2.2 (2), we may allow f to be two-valued at the common endpoints b_1, ..., b_s of consecutive intervals J_i, J_{i+1}.

(b_1) $(J_1, \ldots, J_N)$ *is generating.*

(b_2) X *is an interval of* $\mathbb{R}$ *and* f *is piecewise affine with slopes* $\sigma_1, \ldots, \sigma_N$ *such that* $\prod \sigma_i^{m_i} \neq 1$ *whenever the integers* $m_1, \ldots, m_N$ *are* ≥ 0 *and* $\sum m_i > 0$.

(b_3) X *is an interval of* $\mathbb{R}$ *and* f *is of class* C^3 *with negative Schwarzian derivative*[12]:

$$Sf = \frac{f'''}{f'} - \frac{3}{2}\left(\frac{f''}{f'}\right)^2 < 0.$$

Part (a) has been proved above.

In case (b_1), application of Proposition 2.1 changes only a finite number of periodic points, and application of Proposition 2.4 does not change them, hence α induces a bijection of periodic points up to finite sets and preserves the period.

In case (b_2) let $\xi \in \text{Per}^{\pm}(\hat{f}, m)\backslash\widehat{P}_*$. Then f^m maps $\alpha^{-1}\xi$ into itself, and is affine on this interval, with slope $\neq 1$, hence it has a unique fixed point $x \in \text{Per}^{\pm}(f, m)$. Therefore α induces a bijection of $\text{Per}\, f\backslash P_*$ to $\text{Per}\, \hat{f}\backslash\widehat{P}_*$ preserving the period.

To discuss case (b_3) we have to recall some facts about negative Schwarzian derivatives. (The importance of the condition $Sf < 0$ in the present context was first noted by D. Singer; for a discussion of interval maps with $Sf < 0$ see for instance Preston [34], de Melo [28], Martens [26]. One basic fact is that if $Sf_1 < 0$, $Sf_2 < 0$ then $S(f_2 \circ f_1) < 0$; in particular the condition of negative Schwarzian derivative is preserved by iteration.

Suppose now that $Sf < 0$ and that f is strictly increasing on $[a, b]$; one then checks that f' cannot have a local minimum in (a, b). There can thus be at most one repelling fixed point x of f in (a, b) (i.e. $fx = x$ and $f'(x) > 1$). If x is any fixed point of f, either $fu > u$ for all $u \in (a, x)$, or $fu < u$ for all $u \in (x, b)$, or x is repelling. We apply now these results to f^{2m} restricted to an interval $\bigcap_{k=0}^{m-1} f^{-k}J_{\xi(k)}$, where $(\xi_k) \in \text{Per}^{\pm}(\hat{f}, m)$. In the above interval there is at least one fixed point of f^m, and there is at most one repelling fixed point of f^{2m}; the other fixed points of f^{2m} are of the form $\lim_{n\to\infty} f^{2mn}u$, where $0 \leq k < 2m$ and $f^k u$ is arbitrarily close to one of the division points $a_0, \ldots, a_N$. This implies that the periodic orbits of f which are not repelling form

---

[12]This condition can be weakened. We assume as always that f is continuous and strictly monotone on the intervals $J_i = [a_{i-1}, a_i]$. Suppose also that f is continuously differentiable on (a_{i-1}, a_i) and that $|f'|^{-1/2}$ is strictly convex: these conditions are sufficient for part (b_3) of the proposition to hold (see Preston [34]).

a finite set. Therefore α induces a bijection

$$\text{Per } f \backslash \text{finite set} \;\to\; \text{Per } \hat{f} \backslash \text{finite set}$$

and this bijection preserves the period.

THEOREM 5.5. *Let X be a compact subset of $\mathbb{R}$, $f : X \to X$ a piecewise monotone map, $g : X \to \mathbb{C}$ a function of bounded variation, and $(J_1, \ldots, J_N)$ a minimal cover associated with f.*

If S is a representative set of periodic points we define the zeta function

$$\zeta_S(z) = \exp \sum_{m=1}^{\infty} \frac{z^m}{m} \sum_{x \in S \cap \text{Fix } f^m} \prod_{k=0}^{m-1} g\big(f^k x\big).$$

If Per f *is itself a representative set of periodic points we let*

$$\zeta(z) = \exp \sum_{m=1}^{\infty} \frac{z^m}{m} \sum_{x \in \text{Fix } f^m} \prod_{k=0}^{m-1} g\big(f^k x\big).$$

The function $1/\zeta_S(z)$, or $1/\zeta(z)$, is then analytic for $|z| < \Theta^{-1}$ and its zeros there are the inverses λ^{-1} of the eigenvalues λ of the transfer operator $\mathcal{L}$, with the same multiplicity[13].

[In the case (b_1) of Proposition 5.4 where $(J_1, \ldots, J_N)$ is generating we recover the theorem of Baladi-Keller. In the cases (b_2) of a piecewise affine map and (b_3) of a map with negative Schwarzian derivative we may have attracting as well as repelling periodic orbits].

To prove the theorem we apply successively Propositions 2.1, 2.4, 2.3 and see how they modify the zeta function ζ_S, the parameter Θ, and the transfer operator $\mathcal{L}$.

If we change S by one periodic orbit, of period p through x_p, $1/\zeta_S(z)$ is changed by

$$\exp - \sum_{n=1}^{\infty} \frac{z^{np}}{np} \cdot p \left(\prod_{k=0}^{p-1} g\big(f^k x_p\big) \right)^n = \exp - \sum_{n=1}^{\infty} \frac{1}{n} \left(z^p \prod_{k=0}^{p-1} g\big(f^k x_p\big) \right)^n$$

$$= 1 - z^p \prod_{k=0}^{p-1} g\big(f^k x_p\big)$$

which is holomorphic and without zero for $|z| < \Theta^{-1}$. For the purposes of the theorem we may thus change S by a finite number of periodic orbits. Since S is a representative set of periodic orbits we may, by application of Propositions 2.1 and 2.4 replace ζ_S by the function ζ corresponding to a system with generating partition (we define $\hat{g}$ in

---

[13]The multiplicity of an eigenvalue λ of the operator $\mathcal{L}$ is the dimension of the corresponding generalized eigenspace.

the second step in accordance with the proof of Proposition 3.2 so that $\hat{g}(\alpha x) = g(x)$ if $x \in S/$finite set). Application of Proposition 2.3 does not modify ζ.

Application of Proposition 2.1, 2.4, and 2.3 can only decrease Θ (Proposition 3.2).

Finally, by Proposition 4.3, the transfer operators obtained by successive applications of Propositions 2.1, 2.4, and 2.3 are all Θ-equivalent, and have thus the same eigenvalues λ with the same multiplicity for $|\lambda| > \Theta$.

We have thus reduced the situation of Theorem 5.5 to that of Proposition 5.1, concluding the proof.

COROLLARY 5.6. [14] *In the situation of the theorem let X be an interval of $\mathbb{R}$ and $J_i = [a_i, \alpha_{i+1}]$ for $i = 1, \ldots, N$. Define $\varepsilon : X \to \{-1, 0, +1\}$ such that $\varepsilon(a_0) = \cdots = \varepsilon(a_N) = 0$, and ε is ± 1 on (a_{i-1}, a_i) depending on whether f is increasing or decreasing on that interval. Define the negative zeta function*

$$\zeta^-(z) = \exp 2 \sum_{m=1}^{\infty} \frac{z^m}{m} \sum_{x \in \text{Fix}^- f^m} \prod_{k=0}^{m-1} g\left(f^k x\right)$$

where $\text{Fix}^- f^m = \left\{x \in \text{Fix} f^m : \prod_{k=0}^{m-1} \varepsilon\left(f^k x\right) = -1\right\}$. Then $\zeta^-(z)$ is meromorphic for $|z| < \Theta^{-1}$ and its order[15] at λ^{-1} is $n^\varepsilon(\lambda) - n(\lambda)$ where $n(\lambda)$ and $n^\varepsilon(\lambda)$ are the multiplicities of λ as eigenvalue of $\mathcal{L} = \mathcal{L}_g$ and $\mathcal{L}^\varepsilon = \mathcal{L}_{\varepsilon g}$ respectively.

Since X is an interval, there exists a representative set S of periodic points (Proposition 5.4 (a)). According to Section 5 (and Corollary 2.6) S contains, up to a finite set, the set

$$\bigcup_m \text{Per}^-(f, m) = \bigcup_m \text{Fix}^-(f, m)$$

of negative periodic points. For the analyticity properties of ζ^- we may ignore the finite set and write

$$2 \sum_{x \in \text{Fix}^- f^m} \prod_{k=0}^{m-1} g\left(f^k x\right) = 2 \sum_{x \in S \cap \text{Fix}^- f^m} \prod_{k=0}^{m-1} g\left(f^k x\right)$$

$$= \sum_{x \in S \cap \text{Fix} f^m} \left(\prod_{k=0}^{m-1} g\left(f^k x\right) - \prod_{k=0}^{m-1} \varepsilon\left(f^k x\right) g\left(f^k x\right)\right)$$

---

[14]This is a special case of a conjecture made in Ruelle [**42**] and partially proved in Baladi and Ruelle [**6**].

[15]We define the order of the meromorphic function ζ at z_0 to be the unique $n \in \mathbb{Z}$ such that $(z - z_0)^{-n}\zeta(z)$ is holomorphic and non zero at z_0.

so that

$$\zeta^-(z) = \zeta_S(z)/\zeta_S^\varepsilon(z)$$

where ζ_S^ε is computed from εg instead of g. From this the corollary immediately results.

6. Thermodynamic formalism.

THEOREM 6.1. *Let X be a compact subset of $\mathbb{R}$, f a piecewise monotone map, g a function of bounded variation, and $(J_1, \ldots, J_N)$ a minimal cover associated with f.*

If $g \geq 0$, $R = \max\big(\Theta, \exp P(\log g)\big)$ is the spectral radius of $\mathcal{L}$ acting on $\mathcal{B}$, where

$$P(\log g) = \sup_{\rho \in \mathcal{I}}\big(h(\rho) + \rho(\log g)\big)$$

($\mathcal{I}$ is the set of f-invariant probability measures on X, and $h(\rho)$ the entropy of $\rho \in \mathcal{I}$).

By Theorem 4.1, we know that the spectral radius of $\mathcal{L}$ is

$$R = \lim_{m \to \infty}\big(\|\mathcal{L}^m 1\|_0\big)^{1/m} \geq \Theta.$$

Therefore what we have to prove is that if $\max\big(R, \exp P(\log g)\big) > \Theta$ then $R = \exp P(\log g)$. Furthermore we shall assume $\Theta > 0$ because $\Theta = 0$ implies $R = \exp P(\log g) = 0$.

We consider now several special cases leading to the general situation.

CASE A (FULL SHIFT, CONTINUOUS $g > 0$). The "full shift" hypothesis means that $(J_1, \ldots, J_N)$ is a generating partition with $fJ_i = X$ for $i = 1, \ldots, N$. By standard theory (see Walters [49], Ruelle [35]) we have then

$$\exp P(\log g) = \lim_{m \to \infty}\left(\sum_{x \in \mathrm{Fix}\, f^m} \prod_{k=0}^{m-1} g\big(f^k x\big)\right)^{1/m}.$$

Since $g > 0$ the right-hand side is also the inverse r^{-1} of the radius of convergence r of

$$\zeta(z) = \exp \sum_{m=0}^{\infty} \frac{z^m}{m} \sum_{x \in \mathrm{Fix}\, f^m} \prod_{k=0}^{m-1} g\big(f^k x\big).$$

Theorem 5.5 and Theorem 4.1 (c) imply that if $r > \Theta^{-1}$ then $r = R^{-1}$, and if $r = \Theta^{-1}$ then $R = \Theta$ so that in all cases $R = \exp P(\log g) \geq \Theta$.

CASE B (GENERATING MARKOV PARTITION, $g \geq 0$). Here and in what follows "Markov" means $fJ_i = X$ for $i = 1, \ldots, N$. Using Proposition 2.7 we construct $(\widehat{X}, \hat{f}, \hat{g})$ such that $\hat{g}$ is continuous at periodic points. Notice that X is a Cantor set, and therefore also $\widehat{X}$ is a Cantor set. Let $\bar{g}$ be the smallest u.s.c. function $\geq \hat{g}$; the difference $\bar{g} - \hat{g}$ vanishes outside a countable set which contains no periodic point; furthermore $\sum_{x \in X} \bar{g}(x) - \hat{g}(x) < +\infty$.

Let (A_n) be a decreasing sequence of continuous functions > 0 tending to $\log \bar{g}$. We have, with obvious notation,

$$\Theta_n \searrow \bar{\Theta} = \widehat{\Theta} \leq \Theta$$

where the first step follows from the definition of Θ, the second form Proposition 3.1 and the third from Proposition 3.2. Let us show that also

$$R_n \searrow \bar{R} = \widehat{R}, \quad R = \max(\widehat{R}, \Theta).$$

The first step follows from the definition of R, the second from Corollary 4.5; the last equality follows from the Θ equivalence of $\mathcal{L}$ and $\widehat{\mathcal{L}}$ (Proposition 4.3).

If $\exp P(\log g) > \Theta$ we may in the formula

$$P(\log g) = \sup_\rho \big(h(\rho) + \rho(\log g) \big)$$

assume that $h(\rho) > 0$, and also that ρ is nonatomic. Therefore

$$P(A_n) \geq P(\log \bar{g}) = P(\log \hat{g}) = P(\log g) > \Theta.$$

Since $\exp P(A_n) = R_n$ by case (A), we have

$$R_n \searrow \bar{R} = \widehat{R} = R > \Theta.$$

It remains thus to prove $R = \exp P(\log g)$ under the assumption $R > \Theta > 0$, and we already know that

$$\Theta_n \searrow \bar{\Theta} = \widehat{\Theta} \leq \Theta$$
$$R_n \searrow \bar{R} = \widehat{R} = R.$$

Writing $R/\Theta = e^{2\varepsilon}$ we see that for n large enough

$$R_n/\Theta_n \geq R/\Theta e^\varepsilon = e^\varepsilon.$$

Since $R_n = \exp P(A_n)$ by case (A) we have also

$$\hat{h}(\rho_n) + \rho_n(A_n) = P(A_n) \geq \log \Theta_n + \varepsilon$$

for suitable $\rho_n \in \widehat{\mathcal{I}}$, which must in particular satisfy $\hat{h}(\rho_n) \geq \varepsilon$. Taking a subsequence of ρ_n tending vaguely to $\hat{\rho}$ we have (see *Appendix, Section 7 below*)

$$P(A_n) \searrow P(\log \bar{g}) = \hat{h}(\hat{\rho}) + \hat{\rho}(\log \bar{g})$$
$$\hat{h}(\hat{\rho}) \geq \varepsilon.$$

Therefore there is a nonatomic measure $\hat{\rho}$ satisfying the same conditions. In particular

$$P(\log \hat{g}) \leq P(\log \bar{g}) = \hat{h}(\hat{\rho}) + \hat{\rho}(\log \bar{g}) = \hat{h}(\hat{\rho}) + \hat{\rho}(\log \hat{g}) \leq P(\log \hat{g})$$

so that

$$P(\log \bar{g}) = P(\log \hat{g}) = \hat{h}(\hat{\rho}) + \hat{\rho}(\log \hat{g}).$$

We also have

$$\hat{h}(\hat{\rho}) + \hat{\rho}(\log \hat{g}) = h(\hat{\pi}\hat{\rho}) + (\hat{\pi}\hat{\rho})(\log g) \leq P(\log g).$$

Putting these facts together we have

$$P(A_n) \searrow P(\log \bar{g}) = P(\log \hat{g}) \leq P(\log g)$$

so that for large n

$$P(\log g) \geq P(A_n) - \varepsilon = \log R_n - \varepsilon \geq \log R - \varepsilon \log \Theta + \varepsilon.$$

Therefore the sup in the formula

$$P(\log g) = \sup\Big(h(\rho) + \rho(\log g)\Big)$$

can be replaced by a sup over nonatomic measures ρ with $h(\rho) \geq \varepsilon$; ρ is thus of the form $\hat{\pi}\hat{\rho}$ and this shows that

$$P(\log \hat{g}) = P(\log g).$$

Summarizing

$$P(A_n) \searrow P(\log \bar{g}) = P(\log \hat{g}) = P(\log g)$$

and since we already know that $R_n \searrow \bar{R} = \widehat{R} = R$, and $R_n = \exp P(A_n)$, we have finally $R = P(\log g)$.

CASE C (MARKOV PARTITION, $g \geq 0$). We use Proposition 2.4 to relate the Markov partition to a generating Markov partition. We have thus $\Theta \geq \widehat{\Theta}$ and, by Proposition 4.3,

$$R = \max\Big(\widehat{R}, \Theta\Big)$$

where $\widehat{R} = \exp P(\log \hat{g})$ by case (B).

a.   First we make the assumption $\exp P(\log g) > \Theta$. The sup in

$$P(\log g) = \sup\big(h(\rho) + \rho(\log g)\big)$$

may then be restricted to measures ρ with $h(\rho) \geq \varepsilon$ for some $\varepsilon > 0$ (for instance $\varepsilon = \frac{1}{2}(P(\log g) - \log \Theta)$). If σ is an invariant measure carried by $\pi^{-1}\widehat{Y} = Y = \bigcup_\alpha U_\alpha$ (as defined in Proposition 2.4), the countability of $\widehat{Y}$ implies that $\pi\sigma$ is carried by a union of periodic orbits. If $\pi\sigma$ is carried by a single periodic orbit we may suppose that $\rho(U_\alpha) = \frac{1}{n}$ and $f^n U_\alpha \subset U_\alpha$; then $h(\sigma) = 0$ because $f^n|U_\alpha$ is a homeomorphism. More generally $h(\sigma) = 0$ if σ is carried by $Y = \pi^{-1}\widehat{Y}$. We may thus assume that $P(\log g)$ is given by a sup over ρ so that $\rho(Y) = 0$; therefore

$$P(\log \hat{g}) = P(\log g).$$

We have thus

$$\widehat{R} = \exp P(\log \hat{g}) = \exp P(\log g) > \Theta$$

hence $R = \widehat{R} = P(\log g)$.

b.   We now assume that $R > \Theta$, hence $R = \widehat{R} = \exp P(\log \hat{g})$. Writing $R/\Theta = e^{2\varepsilon}$ we have

$$\exp P(\log \hat{g}) = e^{2\varepsilon}\Theta \geq e^{2\varepsilon}\widehat{\Theta}.$$

Therefore the sup in

$$P(\log \hat{g}) = \sup\big(\hat{h}(\hat{\rho}) + \hat{\rho}(\log \hat{g})\big)$$

may be taken over measures $\hat{\rho}$ with $\hat{h}(\hat{\rho}) \geq \varepsilon$, and we may also take $\hat{\rho}$ nonatomic. We have thus $\hat{\rho}(\widehat{Y}) = 0$ and $\hat{\rho} = \pi\rho$ with $\rho \in \mathcal{I}$ and

$$h(\rho) + \rho(\log g) = h(\hat{\rho}) + \hat{\rho}(\log \hat{g}).$$

Therefore $\exp P(\log g) \geq \exp P(\log \hat{g}) = R > \Theta$ and we are brought back to case A.

CASE D (GENERAL PARTITION, $g \geq 0$).   We use Proposition 2.3 to relate the general partition to a Markov partition. We have here $\widehat{\Theta} = \Theta$. Identifying X with a subset of $\widehat{X}$ we know that if $\rho \in \widehat{\mathcal{I}}$ and $\rho(\log \hat{g})$ is finite then $\operatorname{supp} \rho \subset X$. Therefore

$$P(\log \hat{g}) = \sup_{\rho \in \widehat{\mathcal{I}}}\big(h(\rho) + \rho(\log \hat{g})\big)$$

$$= \sup_{\rho \in \mathcal{I}}\big(h(\rho) + \rho(\log g)\big) = P(\log g).$$

If $\exp P(\log g) > \Theta$, hence $P(\log \hat{g}) > \widehat{\Theta}$, we have $\widehat{R} = \exp P(\log \hat{g})$ by case (C), hence $\widehat{R} > \widehat{\Theta}$. Therefore the assumption $\max\big(R, \exp P(\log g)\big)$

$> \Theta$ implies $\max\left(R, \widehat{R}\right) > \Theta = \widehat{\Theta}$ hence, by Proposition 4.3, $R = \widehat{R} > \Theta$, hence $R = \widehat{R} = \exp P(\log \hat{g})$ by case (C), hence $R = \exp P(\log g)$.

CASE E (GENERAL CASE). We use Proposition 2.1 to relate the minimal cover to a partition. We have here $\widehat{\Theta} = \Theta$. If $\exp P(\log g) > \Theta$ the usual arguments give $P(\log g) = P(\log \hat{g})$, and since $\exp P(\log \hat{g}) = \widehat{R}$ by case (D), we have $\widehat{R} > \Theta$. So the assumption $\max\left(R, \exp P(\log g)\right) > \Theta$ implies $\max\left(R, \widehat{R}\right) > \Theta$, hence (using Proposition 4.3) $\widehat{R} = R > \Theta = \widehat{\Theta}$. Since $\widehat{R} = \exp P(\log \hat{g})$ by case (D) we have $\exp P(\log \hat{g}) > \widehat{\Theta}$. Again by the usual arguments we obtain thus $P(\log g) = P(\log \hat{g})$ so that $R = \widehat{R} = \exp P(\log \hat{g}) = \exp P(\log g)$ which concludes the proof.

THEOREM 6.2. [16] *If $g \geq 0$ and $\exp P(\log g) > \Theta$, then the set of equilibrium states*

$$\Delta = \left\{\rho \in \mathcal{I} : h(\rho) + \rho(\log g) = P(\log g)\right\}$$

is non empty; Δ is a simplex and its vertices are ergodic measures with entropy $h \geq P(\log g) - \log \Theta$.

(We make the usual assumptions on X, f, g, in particular g has bounded variation).

If the entropy h is u.s.c. on $\mathcal{I}$, and g u.s.c. on X then

$$\rho \mapsto h(\rho) + \rho(\log g)$$

is affine u.s.c. on $\mathcal{I}$, hence reaches its maximum $P(\log g)$ on a *face* Δ of the Choquet simplex $\mathcal{I}$, and Δ is also a simplex. Furthermore the vertices (= extremal points) of Δ are ergodic measures because Δ is a face of $\mathcal{I}$.

In the proof of Theorem 6.1 we have reduced the general case to that of a system $(\widehat{X}, \hat{f}, \bar{g})$ (see part (b) of 6.1) where $\hat{h}$ and $\bar{g}$ are u.s.c. The set

$$\bar{\Delta} = \left\{\rho \in \widehat{\mathcal{I}} : \hat{h}(\rho) + \rho(\log \bar{g}) = P(\log \bar{g})\right\}$$

is thus a face of $\mathcal{I}$, a simplex, and its vertices are ergodic measures. Since $\rho(\log \bar{g}) \leq \log \widehat{\Theta}$ we have $\hat{h}(\rho) \geq P(\log \bar{g}) - \log \widehat{\Theta} > 0$ for $\rho \in \bar{\Delta}$, hence $\bar{\Delta}$ consists of nonatomic measures. We use the vague topology on $\widehat{\mathcal{I}}$ and $\bar{\Delta}$, i.e., the topology of pointwise convergence on the space $\widehat{\mathcal{C}}$ of continuous functions $\widehat{X} \to \mathbb{C}$. We do not change this topology on $\bar{\Delta}$ if we replace $\widehat{\mathcal{C}}$ by $\widehat{\mathcal{C}} \cup \widehat{\mathcal{B}}$ where $\widehat{\mathcal{B}}$ is the space of functions $\widehat{X} \to \mathbb{C}$ with bounded variation (elements of $\widehat{\mathcal{B}}$ define continuous functions on $\bar{\Delta}$ because $\bar{\Delta}$ consists of nonatomic measures). Again we do not

---

[16]The proof of this theorem uses Choquet theory (simplexes, faces, etc.) for which see for instance Choquet and Meyer [10].

change the topology on $\bar{\Delta}$ if we replace $\widehat{\mathcal{C}} \cup \widehat{\mathcal{B}}$ by $\widehat{\mathcal{B}}$ (this gives a weaker Hansdorff topology, therefore equivalent to the original topology). In conclusion the vague topology of $\bar{\Delta}$ is also the topology of pointwise convergence on $\widehat{\mathcal{B}}$. The constructions in the proof of Theorem 6.1 give a linear homeomorphism $\Delta \to \Delta$ where the topology on Δ is that of pointwise convergence on $\mathcal{B}$, i.e., also the vague topology. Therefore Δ is not empty, it is a simplex, and its vertices are ergodic measures with entropy $h \geq P(\log g) - \log \Theta$.

7. Appendix: Extension of the definition of pressure.

Let X be compact metrizable, and $f : X \to X$ continuous. We say that f is *expansive* if, given an allowed metric d, there is $\varepsilon > 0$ such that $d(f^k x, f^k y) \leq \varepsilon$ for all $k \geq 0$ implies $x = y$. With the terminology of Chapter 1, we see in particular that if $X \subset \mathbb{R}$ and (X, f) has a generating partition, then f is expansive.

We denote by $\mathcal{I}$ the set of f-invariant probability measures on M, with the vague topology (such that $\mathcal{I}$ is compact). If $\rho \in \mathcal{I}$, the entropy $h(\rho)$ is either ≥ 0 or $+\infty$. The function $h(\cdot)$ is affine, and if f is expansive, $h(\cdot)$ is finite and u.s.c. (upper semicontinuous). [See Walters [49], Ruelle [39], and note that the inverse limit construction allows one to pass from a map (an expansive map) to a homeomorphism (an expansive homeomorphism, defined so that $d(f^k x, f^k y) \leq \varepsilon$ for all $k \in \mathbb{Z}$ implies $x = y$)].

For continuous $A : X \to \mathbb{R}$, the *pressure* $P(A)$ is defined by

$$P(A) = \sup_{\rho \in \mathcal{I}} \Big(h(\rho) + \rho(A) \Big).$$

[There is another definition of the pressure, and its equivalence with the above formula is Walters' [49] theorem. The concept of pressure comes from statistical mechanics, see Ruelle [35]]. We say that $\rho \in \mathcal{I}$ is an *equilibrium state* if $h(\rho) + \rho(A) = P(A)$. In particular when $h(\cdot)$ is u.s.c. there is always at least one equilibrium state for A.

One can also define the pressure (by the above formula) and equilibrium states for some non continuous A. We now indicate a case where this extension is useful.

If h is finite u.s.c. on $\mathcal{I}$ and g u.s.c. on M, $g \geq 0$, then

(7.1) $$\rho \mapsto h(\rho) + \rho(\log g)$$

is affine u.s.c. : $\mathcal{I} \mapsto \mathbb{R} \cup \{-\infty\}$ and one may define

$$P(\log g) = \max_{\rho \in \mathcal{I}} \Big(h(\rho) + \rho(\log g) \Big).$$

If (A_n) is a decreasing sequence of continuous functions tending to $\log g$, then $P(A_n) \to P(\log g)$. If ρ_n is an equilibrium state for A_n, and $\rho_n \to \rho$ (vaguely), then ρ is an equilibrium state for $\log g$.

[Note that $\rho \mapsto \rho(\log g)$ is limit of the decreasing sequence of continuous functions $\rho \mapsto \rho(A_n)$ hence u.s.c., so that (7.1) is also u.s.c. Since

$$P(A_n) = h(\rho_n) + \rho_n(A_n)$$

and $\rho_n \to \rho$, we have

$$P(\log g) \leq \lim_{n\to\infty} P(A_n) \leq h(\rho) + \rho(A_n) \to h(\rho) + \rho(\log g) \leq P(\log g)$$

hence $P(A_n) \to P(\log g)$ and ρ is an equilibrium state for $\log g$].

Bibliography

1. M. Artin and B. Mazur, *On periodic points*, Ann. of Math. (2) **81** (1965), 82–99.
2. M. Atiyah and R. Bott, *A Lefschetz fixed point formula for elliptic complexes*, Ann. of Math. **86** (1967), 374–407; **88** (1968), 451–491.
3. V. Baladi, *Dynamical zeta functions*, Real and Complex Dynamical Systems (B. Branner and P. Hjorth, eds.), Kluwer Academic Publishers (to be published).
4. V. Baladi and G. Keller, *Zeta functions and transfer operators for piecewise monotone transformations*, Comm. Math. Phys. **127** (1990), 459–477.
5. V. Baladi and D. Ruelle, *An extension of the theorem of Milnor and Thurston on zeta functions of interval maps*, Ergodic Theory Dynamical Systems (to appear).
6. ———, *Some properties of zeta functions associated with maps in one dimension* (in preparation).
7. P. Billingsley, *Ergodic Theory and Information*, John Wiley, New York, 1965.
8. R. Bowen, *Equilibrium states and the ergodic theory of Anosov diffeomorphisms*, Lecture Notes in Math. vol. 470, Springer-Verlag, Berlin, 1975.
9. R. Bowen and O.E. Lanford, *Zeta functions of restrictions of the shift transformation*, Global Analysis, Proc. Symp. Pure Math. vol. 14, Amer. Math. Soc., Providence, R.I. (1975), pp. 43–49.
10. G. Choquet and P.-A. Meyer, *Existence et unicité des représentations intégrales dans les convexes compacts quelconques*, Ann. Inst. Fourier (Grenoble) **13** (1963), 139–154.
11. M. Denker, C. Grillenberger and K. Sigmund, *Ergodic theory on compact spaces*, Lecture Notes in Math. vol. 527, Springer-Verlag, Berlin, 1976.
12. D. Fried, *The zeta functions of Ruelle and Selberg* I, Ann. Sci. École Norm. Sup. (4) **19** (1986), 491–517.
13. ———, *Rationality for isolated expansive sets*, Adv. in Math. **65** (1987), 35–38.
14. ———, *The flat-trace asymptotics of a uniform system of contractions* (Preprint).
15. A. Grothendieck, *Produits tensoriels topologiques et espaces nucléaires*, Mem. Amer. Math. Soc. vol. 16, Providence, R.I, 1955.
16. ———, *La théorie de Fredholm*, Bull. Soc. Math. France **84** (1956), 319–384.
17. J. Guckenheimer, *Axiom A + no cycles $\Rightarrow \zeta_f(t)$ rational*, Bull. Amer. Math. Soc. **76** (1970), 592–594.
18. V. Guillemin and Sh. Sternberg, *Geometric asymptotics*, Math. Surveys vol. 14, Amer. Math. Soc., Providence, R.I., 1977.
19. N. Haydn, *Meromorphic extension of the zeta function for Axiom A flows*, Ergodic Theory Dynamical Systems **10** (1990), 347–360.
20. F. Hofbauer, *Piecewise invertible dynamical systems*, Probab. Theor. Relat. Fields **72** (1986), 359–386.
21. F. Hofbauer and G. Keller, *Zeta-functions and transfer-operators for piecewise linear transformations*, J. Reine Angew. Math. **352** (1984), 100–113.
22. G. Keller and T. Nowicki, *Spectral theory, zeta functions and the distribution of periodic points for Collet-Eckmann maps*, Comm. Math. Phys. **149** (1992), 31–69.
23. G. Levin, M. Sodin and P. Yuditskii, *A Ruelle operator for a real Julia set*, Comm. Math. Phys. **141** (1991), 119–131.
24. ———, *Ruelle operators with rational weights for Julia sets*, J. Analyse Math. (to appear).
25. A. Manning, *Axiom A diffeomorphisms have rational zeta functions*, Bull. London Math. Soc. **3** (1971), 215–220.
26. M. Martens, *Interval dynamics*, Thesis, Delft, 1990.

27. D. Mayer, *Continued fractions and related transformations*, Ergodic Theory, Symbolic Dynamics and Hyperbolic Spaces (T. Bedford, M. Keane, C. Series, eds.) Oxford University Press, Oxford, 1991.

28. W. de Melo, *Lectures on one-dimensional dynamics*, 17e Colóquio Brasileiro de Matemática, Rio de Janeiro.

29. J. Milnor and W. Thurston, *On iterated maps of the interval*, Dynamical Systems, Lecture Notes in Mathematics vol. 1342, Springer, Berlin, 1988, pp. 465–563.

30. Nihon Sugakkai, ed., *Encyclopedic Dictionary of Mathematics*, MIT Press, Cambridge, Mass., 1977.

31. R.D. Nussbaum, *The radius of the essential spectrum*, Duke Math. J. **37** (1970), 473–478.

32. W. Parry and M. Pollicott, *An analogue of the prime number theorem for closed orbits of Axiom A flows*, Ann. of Math. (2) **118** (1983), 573–591.

33. _____, *Zeta Functions and the Periodic Orbit Structure of Hyperbolic Dynamics*, Société Mathématique de France (Astérisque vol. 187–188), Paris, 1990.

34. C.J. Preston, *Iterates of maps on an interval*, Lecture Notes in Mathematics vol. 999, Springer, Berlin, 1983.

35. D. Ruelle, *Statistical mechanics on a compact set with Z^ν action satisfying expansiveness and specification*, Bull. Amer. Math. Soc. **78** (1972), 988–991; Trans. AMS **185** (1973), 237–251.

36. _____, *Zeta functions and statistical mechanics*, Astérisque **40** (1976), 167–176.

37. _____, *Generalized zeta-functions for axiom A basic sets*, Bull. Amer. Math. Soc. **82** (1976), 153–156.

38. _____, *Zeta-functions for expanding maps and Anosov flows*, Invent. Math. **34** (1976), 231–242.

39. _____, *Thermodynamic Formalism*, Addison-Wesley, Reading MA, 1978.

40. _____, *The thermodynamic formalism for expanding maps*, Comm. Math. Phys. **125** (1989), 239–262.

41. _____, *An extension of the theory of Fredholm determinants*, Inst. Hautes Études Sci. Publ. Math. **72** (1990), 175–193.

42. _____, *Spectral properties of a class of operators associated with maps in one dimension*, Ergodic Theory Dynamical Systems **11** (1991), 757–767.

43. _____, *Analytic completion for dynamical zeta functions*, Helv. Phys. Acta **66** (1993), 181–191.

44. _____, *Functional equation for dynamical zeta functions of Milnor-Thurston type* (to appear).

45. H.H. Rugh, *The correlation spectrum for hyperbolic analytic maps*, Nonlinearity **5** (1992), 1237–1263.

46. S. Smale, *Differentiable dynamical systems*, Bull. Amer. Math. Soc. **73** (1967), 747–817.

47. F. Tangerman, *Meromorphic continuation of Ruelle zeta functions*, Boston University thesis, 1986 (unpublished).

48. P. Walters, *Ergodic Theory. Introductory Lectures*, Lecture Notes in Math. vol. 458, Springer-Verlag, Berlin, 1975.

49. _____, *A variational principle for the pressure of continuous transformations*, Amer. J. Math. **97** (1976), 937–971.